A NATURALIST'S GUIDE TO THE

BIRDS
OF
BRITAIN
and Northern Europe

A NATURALIST'S GUIDE TO THE

BIRDS
OF
BRITAIN
and Northern Europe

Peter Goodfellow & Paul Sterry

JOHN BEAUFOY PUBLISHING

This edition first published in the United Kingdom in 2019 by
John Beaufoy Publishing Limited
11 Blenheim Court, 316 Woodstock Road, Oxford OX2 7NS, England
www.johnbeaufoy.com

10 9 8 7 6 5 4 3 2 1

ISBN 978-1-912081-21-9

PICTURE CREDITS

Front cover: *Main image:* Linnet; *bottom left:* Curlew: *bottom centre:* Slavonian
Grebe; *bottom right:* Lapwing. **Back cover:** Bullfinch. **Title page:** Great Spotted
Woodpecker. **Contents page:** Sparrowhawk. All by Paul Sterry.
Main descriptions: all photographs by Paul Sterry except for those on pages
92(top) and 150(top), which were supplied by Brian E. Small, and the following,
which were supplied by Nature Photographers Ltd:
Frank Blackburn 97(top), 99(top); Mark Bolton 83(top), 117(bottom); Laurie
Campbell 14(top), 40(top); Alan Drewitt 49(bottom); Ernie Janes 45(bottom),
90(top), 92(bottom); Andrew Merrick 77(bottom), 108(bottom); Owen Newman
46 (top); Philip Newman 42(bottom), 48(bottom left), 50(bottom); David Osborn
20(top), 39(bottom), 47(bottom); Bill Paton 51(bottom); Don Smith 91(bottom),
134(top); E.K. Thompson 89(bottom); Roger Tidman 39(top), 42(top left), 43(top
& bottom), 48(top), 51(top), 70(top), 73(bottom), 81(top), 91(top), 101(bottom),
122(bottom), 135(top), 152(top), 153(top); Derek Washington 82(top).

DEDICATION
To June, who has looked after me up hill and down dale.

Edited, designed and typeset by D & N Publishing, Baydon, Wiltshire, UK

Printed and bound in Malaysia by Times Offset (M) Sdn. Bhd.

·CONTENTS·

ACKNOWLEDGEMENTS

No work of this kind can be the sole work of the author. First, I am very grateful to John Beaufoy and D & N Publishing for getting me started; second, to Paul Sterry, who has provided all the illustrations, without which your efforts to identify the birds listed here would be made very much harder; and, finally, to the many birdwatchers who have provided so much information in recent years on identification, distribution and behaviour.

INTRODUCTION

Many people who would not go as far as calling themselves birdwatchers are nevertheless really interested in birds. When they are out and about in the countryside and see a flash of colourful plumage, hear a strange call or witness a striking bit of behaviour, they may well stop and say to themselves, 'I wish I knew what that bird was!' It is hoped that this book will solve that problem, acting as an introductory guide to help residents, travellers and visitors across the region identify the birds they are most likely to see and hear, at any time of the year. In all, more than 280 species have been included. They represent a wide variety of breeding species, plus others that breed outside the area covered by this book but visit these shores on migration in spring and autumn.

For each species covered, the English common name and scientific name are provided, along with measurements of its length (L) and wingspan (WS). Also included are one or two photographs, and a paragraph on each of description, distribution, and habits and habitat (see also 'Bird Identification', p. 10). Unless otherwise stated, the photographs show adult birds; these are adult males when there is only one photograph for species that have distinctive male and female plumages, and are in summer plumage if there are distinctive summer and winter plumages. Exceptions to this are labelled. Differences in plumage between the sexes, and at various ages and seasons, are explained in the 'Description' section in the text. The 'Distribution' section names the countries in which the bird may be found, e.g. from the British Isles eastwards to Denmark – the countries in between are not named. This section also states whether the bird is resident, a summer visitor to breed and where it winters elsewhere, or whether it nests outside our region and so is seen only on migration. An observer needs to read the 'Habits and habitat' section as well, to narrow down the part of the country concerned; some general knowledge or further research will be needed to find exactly where that is. The comments on the birds' habits are selective – one comment may emphasise feeding habits (as with the terns) or another something special about breeding (as with the Cuckoo), but most give a description of the bird's voice.

Geography

Scientists have long divided the world into six biogeographic regions, each with a noticeably different fauna. The six regions are the Nearctic, Palaearctic, Neotropic, Afrotropic, Orient and Australasia, within which the animals are grouped further by the particular habitats they occupy. The Western Palaearctic is a region in which the families of birds have much in common. It has rather irregular boundaries, but includes all of Europe from Iceland eastwards to the Ural Mountains (80°E) and southwards to North Africa, including Libya and Egypt; it then sweeps northwards via Iraq to the west and north banks of the Caspian Sea, from whence it rejoins the Urals.

The area covered by this book is the northwestern part of the Western Palaearctic, which roughly speaking embraces all the land within the region northwards from about 45°N and eastwards to the Kola Peninsula at 40°E – i.e. Iceland, the British Isles (United Kingdom and Eire), northern France, Belgium, the Netherlands, Germany, Poland, the Czech and Slovak republics, Denmark, Sweden, Norway, Finland, the Baltic states (Lithuania, Latvia and Estonia), Belarus and just into Russia.

Climate

In general, the region's climate ranges from Arctic, through boreal (sub-Arctic) to temperate zones. The powerful Gulf Stream current carries warm water across the Atlantic from the Gulf of Mexico to western Europe and, together with the prevailing westerly winds that are warmed by the sea, gives this region milder weather than places in Asia and North America at similar latitudes. For example, in January, Berlin in Germany is about 8°C warmer than Calgary in Canada. The most spectacular effects of the Gulf Stream are perhaps seen along the Norwegian coast, much of which lies within the Arctic and yet almost all of which is ice-free in winter, when the interior is covered with snow and ice. Further south, we find that Glasgow in Scotland has an average temperature of +3°C in January, yet in Moscow, Russia, on about the same latitude, it drops to -10°C. The Isles of Scilly, at around 49°N and nearly 50km west of the southwestern tip of England, rarely experience frost or snow and have a climate so mild and warm that they are famous for supporting Mediterranean plants – although they do suffer from autumn's westerly gales, much to the delight of birdwatchers (p. 13)!

Habitats

The region covered by this book has a landscape that varies from the flat lowlands of East Anglia and the Netherlands, to the mountainous parts of Scotland (>1,300m) and Fennoscandia (>2,000m). In the far north, the habitat is tundra (vast tracts of nearly level, treeless ground), which merges into boreal forest of birch, willow and the taiga (coniferous forest), stretching from here right across Asia. Northern mountains and hills are rocky and support specialist birds such as grouse and Snow Buntings.

Much of northwest Europe was deciduous forest in primeval times, but the vast majority of this has since been cleared by Man to create pasture for cattle and sheep, and arable

Many sea cliffs of the N and W have internationally important seabird populations.

land for crops. Deforestation and the destruction of other natural habitats has greatly changed – and is still changing – the region's bird life. Isolated areas of ancient forest, moorland, lowland heath, wetlands and grasslands now often exist only in national parks and nature reserves.

An appreciation of different habitats is important in order to see as many of northern Europe's birds as possible. Some birds, such as Wrens, will live in a wide range of habitats, but many families and individual species usually breed or winter in a particular habitat – gulls on a sea cliff, Bearded Tits in a reedbed, Jays in woodland, Stonechats on heath, Red Grouse on heather moors, and so on. Many waders can be observed only if you visit coastal or estuarine habitats in spring or autumn. Some species, like House Sparrows and other garden birds seen at bird feeders, live comfortably with Man; others are very wary of Man, such as some birds of prey, and need time and patience to be found. Many habitats are also quite different in winter and summer – a summer estuary with a few Herring Gulls, a Grey Heron and a pair of Shelducks may have a winter population of hundreds of gulls, waders and wildfowl, each of several different species; a spring wood full of the song of tits and summer visitors will be all but silent in winter. The best advice is to visit your local patch regularly; by becoming familiar with its birds and their habitat preferences, you can more confidently head further afield.

Wetland: both tidal saltmarsh and flooded water meadows, are important wintering grounds for wildfowl and waders.

Reedbeds everywhere are in danger of being destroyed – drained for farmland or housing developments.

Tundra may seem desolate and treeless, but many species breed there in the short summer.

Deciduous woodland is perhaps the richest habitat in the region for its range of species.

Taxonomy and Nomenclature

The classification of birds has taxed the minds of scientists for centuries. Skeletal structure, plumage, voice and behaviour have all helped give order to the 10,000 or so species found worldwide. Nowadays, DNA testing is revealing a new understanding of what a species is and where it sits in relation to other birds. For several hundred years, birds and other living organisms have been allocated names derived from Latin or Greek words. Swedish scientist Carl von Linné (or Linnaeus, to use his Latinised name) standardised the naming of species in his book *Systema Naturae*, published in 1735, adopting a neat and tidy binomial system instead of the many words some scientists were using. So, the House Sparrow became *Passer* (family name or genus) *domesticus* (personal or specific name). Later, it was found that an identifiable geographic variation of a species needed the addition of a third name to distinguish it. For example, the Yellow Wagtail of the British Isles became *Motacilla flava flavissima* to identify it clearly – Linnaeus first classified and named the bird in 1735, and British zoologist Edward Blyth reclassified it as a subspecies in 1834, giving it its third name. These scientific names are understood as belonging to a single species across the world, no matter what language the observer may speak. Thus, the scientific name *Passer domesticus* is the same species, whether it is called locally the House Sparrow (English), Moineau domestique (French), Huismus (Dutch), Haussperling (German), Gråsparv (Swedish) or Домовый воробей (Russian).

The order in which zoologists place birds is one with which you will become more familiar as you use the book more. Roughly speaking, it starts with the most ancient, primitive birds (divers and petrels) and ends with the species that have evolved most recently (passerines, i.e. the songbirds). This orderly arrangement is constantly under review, so may differ slightly from one book to another. This guide follows the much-used classification devised in 1977 by Dutch ornithologist K.H. Voous and used in the definitive book *The Birds of the Western Palaearctic* by S. Cramp and C.M. Perrins. However, a few species have had to be moved slightly from their official order owing to the constraints of layout; even so, such species are still placed close to their nearest relatives.

Bird Identification

When attempting to identify a 'new' species, there is a great danger for the observer to focus on one or two things he or she has seen rather than observing the whole bird. For example, if you spot a bird with a red head in the garden one morning you will not be able to identify it without noting more detail – depending on where you are, it could be a Red-headed Bunting, a crossbill, a Scarlet Rosefinch or one of several other species. The illustrations below shows the names of the parts of a bird that are commonly used to describe its appearance. Do become familiar with them. A longer, closer look at our garden bird may show that its forehead is red, with bands of white and black behind that, and when it flies, a broad yellow bar stretches across its secondaries and primaries. With this information it can be narrowed down to a single species, as the only bird that fits this description is a Goldfinch!

Very often, a species can be identified reliably only when the observer hears its call or song. This introductory book to the region's birds does not have the space to include all the species' calls and songs, which in any case is a notoriously difficult task. You will, however, find comments about the sounds many of the birds make, in order to illustrate that special aspect of the bird's habits and, importantly, to aid identification. For example, the leaf warblers (pp. 126–27) can be separated with certainty only when their song is heard (or you have one in the hand!). Try to go out with an experienced 'bird listener', or obtain a set of recordings of bird voices (*see* 'References and Further Reading', p. 156).

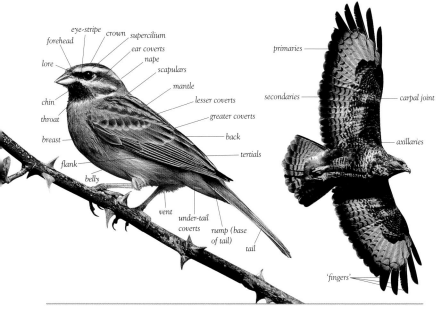

GLOSSARY

For a guide to bird topography, *see* the illustration opposite.

Arthropod Any invertebrate with an exoskeleton (the supporting structure covering the outside of the body in many animals), jointed limbs and a segmented body, such as insects, spiders and centipedes.

Boreal The northern climatic zone of short summers and long, snowy winters.

Broadleaved Trees that are not coniferous and shed their leaves in the autumn.

Carr An area of bog in which scrub, especially willows or alders, has become established.

Coniferous Trees such as pines, firs, larches, spruces, cedars, junipers and yews, which bear seeds in cones.

Conspecific Belonging to the same species.

Continental shelf The seabed surrounding a continent at depths up to *c.* 220m.

Continental slope The edge of the continental shelf, where it drops off steeply.

Coppice A dense growth of small trees, regularly trimmed back to stumps to encourage new growth.

Crepuscular Active at twilight or just before dawn.

Deciduous Trees that shed their leaves each year in the autumn.

Defoliating Depriving a plant or tree of its leaves.

Diurnal Active during the day.

Drumming The sound made in flight by the outer tail feathers of Snipe (p. 71) as it dives in courtship display.

Eyrie The nest of an eagle or other bird of prey.

Fennoscandia The countries of Norway, Sweden and Finland.

Feral Existing in the wild after having been domesticated and then escaped or released.

Gamebird A bird hunted for sport, e.g. grouse, partridges, Pheasant.

Insectivorous Eats mainly insects, including their larvae and pupae.

Invertebrate Any animal lacking a backbone.

Irruption The sudden entry to a region of large numbers of birds.

Juvenile Young birds that have fledged but have not yet moulted in the autumn into their first adult plumage.

Lagoon A body of water cut off from the sea by sand bars.

Lek The communal display of cock birds to hens gathered around them.

Molluscs Animals belonging to the phylum Mollusca, which includes land and sea snails and slugs, and bivalves (shellfish with two shells, such as mussels, cockles and clams).

Nocturnal Active at night.

Omnivorous Eats both plants and animals.

Passage migrant Does not breed in the region but is seen there on its way to breeding or wintering grounds.

Pelagic Lives or occurs on the open sea, not normally coming to land except to breed.

Plankton Small, drifting animals and plants living in the surface layer of the sea.

Pollarded Trees that have had their branches cut back to encourage more growth, e.g. as with willows and Hazel *Corylus avellana*.

Polyandrous When the female of a species mates with more than one male during the breeding season.

Polygynous The practice of a male mating with more than one female.

Raptor A diurnal bird of prey, such as the eagles, hawks and falcons.

Resident Birds that do not migrate to or from the region in which they breed.

Roding The aerial display of Woodcock (p. 70).

Sahel A vast semi-arid region of North Africa south of the Sahara, forming a transition zone between the desert and savannah.

Saltmarsh An area of marshy ground that is intermittently inundated with salt water by the tides.

Savannah Open grassland with scattered bushes or trees, characteristic of much of tropical Africa.

Scandinavia The countries of Denmark, Sweden and Norway.

Seabird A general term for coastal and pelagic birds such as gulls, terns, shearwaters, petrels, Gannet and Cormorant.

Sedentary Birds that are resident in a certain region and rarely or never move from the territory in which they nest.

Speculum A patch of distinctive, often diagnostic colour on the inner secondaries of dabbling ducks.

Sphagnum bog Any bog in temperate climates with sphagnum mosses as the main vegetation; the vegetation decays to form peat.

Spitzbergen One of the islands of the Norwegian archipelago of Svalbard, sometimes used to describe the whole group.

Steppe Extensive grassy plains, usually without trees.

Summer visitor A bird that migrates (mostly from the south) to breed in the summer, then returns in winter from whence it came.

Taiga Coniferous forest across much of the sub-Arctic, bordered by tundra to the north and steppe to the south.

Tundra A vast treeless zone lying between the Arctic ice cap and the taiga, with a permanently frozen subsoil. It also occurs in the Antarctic.

Vagrant A rare, accidental visitor from another region, usually at migration time.

Wader Several closely related families of shore birds, usually with a long bill and legs, which feed in mud at the water's edge or on farmland, e.g. sandpipers, plovers, Curlew.

Winter visitor A bird that migrates to spend the winter (mostly from the north and east), then returns from whence it came to breed.

Wreck The sudden arrival of exhausted seabirds such as petrels and Little Auk, close to or on land, after prolonged, severe gales at sea.

WHERE TO WATCH BIRDS IN NORTHERN EUROPE

In a region as large as that covered by this book, there are bound to be places that are particularly good for birdwatching. Travelling birdwatchers are fortunate today in being able to buy field guides to many countries in this region and around the world (*see* p. 156). There is space here to mention only a few of the most famous sites:

BRITISH ISLES
- Isles of Scilly, in spring and autumn, especially the latter for rare migrants. The islands are one of the best places to observe migrants from North America, and in September or October it is even possible to see rare visitors from both North America and Asia on the same day!
- Minsmere, Suffolk, one of Britain's finest reserves, owned by the Royal Society for the Protection of Birds (RSPB). Here you can see over 100 species in a day in May.
- Farne Islands, Northumberland, within easy reach by boat on a day trip from Seahouses. It is home to easily observed seabird colonies of nearly 20 species – a walk among the terns is a must!

FRANCE
- Parc Naturel Régional de Brière (Brière Regional Natural Park), near Nantes, one of the best wetlands in Europe, with a park office on Île de Fedrun. The place to see harriers, rails, Black Terns and much more.
- Fôret d'Argonne (Argonne Forest), east of Reims, best known as a splendid place to find up to 10 species of birds of prey.

GERMANY
- Nationalpark-Wattenmeer (Wadden Sea National Park, a UNESCO site), which takes in all the country's North Sea coast. It is divided into three separate national parks, each with visitor centres.

POLAND
- Puszcza Bialowieska (Bialowieska Forest) for rare owls, birds of prey and woodpeckers.
- Biebrzanski Park Narodowy (Biebrza National Park), whose marshes in spring are home to Common Cranes, Great Snipe and eagles.

FENNOSCANDIA
- Falsterbo Peninsula, Sweden, long famous as a site to watch the autumn migration, especially of over a dozen species of raptors in September.
- Hardangervidda Nasjonalpark (Hardangervidda National Park), Norway, the largest mountain plateau in northern Europe. The breeding season brings wildfowl, Common Cranes, waders (including Temminck's Stint), Bluethroats and buntings.
- Patvinsuon Kansallispuisto (Patvinsuo National Park), Finland, covers thousands of hectares of moorland and ancient forest, and is home to bears, wolves, four species of grouse, waders, owls and Siberian Tits.

Black-throated Diver

▪ *Gavia arctica* L 63–75cm
WS 100–122cm

DESCRIPTION All divers look long-bodied with no tail. Summer adult: pale grey head and nape; black throat outlined with narrow black and white stripes, underparts white. Black dagger-shaped bill. Striking white patches on back. Wings black. Winter adult: upperparts plain dark grey. Plain white neck and underparts; swims low, so latter hardly show save for white patch at rear of flanks. Juvenile: as winter adult but with pale feather edges to upperparts.

DISTRIBUTION Breeds in Iceland and Fennoscandia eastwards, plus a few in N Scotland.
HABITS AND HABITAT Breeds on freshwater lakes with fish. Winters on coasts further S, often in flocks. Legs are set far back, so struggles on land; nest is very close to water's edge. Dives for fish, slipping underwater, not leaping out first like Cormorant (p. 20). Territorial song: loud, repeated *clooee*.

Red-throated Diver

▪ *Gavia stellata* L 55–67cm
WS 91–110cm

DESCRIPTION Summer adult: pale grey head and neck, with rufous-red throat (can look black at a distance). Upperparts plain, dark grey-brown. Winter adult: pale grey crown and hind-neck. Face and rest of underparts white; swims low, so often only white on head and neck can be seen, unless it rolls on its side to preen. Upperparts finely speckled with white. Swims with bill pointing up.

DISTRIBUTION Breeds in Iceland, N Scotland, and Fennoscandia eastwards. Winters, often in loose flocks, around those coasts and S to Spain and the Black Sea.
HABITS AND HABITAT Breeds on small lakes, tundra pools or forest bogs; flies to larger lakes or the coast to fish. Nests at water's edge as it struggles on land. Silent in winter, but on breeding territory pairs duet with loud, far-carrying calls.

Great Northern Diver

■ *Gavia immer* L 73–88cm
WS 122–148cm

DESCRIPTION Has a steep forehead,
unlike its relatives. Summer adult:
black head, neck and wings; white-
striped patch on side of neck, chequered
black and white upperparts. Heavy
black bill held horizontally. Winter
adult: blackish upperparts and white
below; bill greyish white with dark tip.
Immature: pale feather edges on back.
DISTRIBUTION Breeds only in Iceland
in region, but widely in N North
America. Winters on coasts of Iceland, Norway, British Isles and English Channel.
HABITS AND HABITAT Breeds beside deep lakes or wooded bays. As with other
divers, flies with neck outstretched and wings apparently set well back. Dives for fish.
Call: eerie, laughing; often used as atmospheric background soundtrack in films set in
desolate N territories.

Little Grebe

■ *Tachybaptus ruficollis*
L 25–29cm WS 40–45cm

DESCRIPTION Dumpy, with
prominent, blunt-ended body. Summer
adult: chestnut face highlighted by
bright yellow gape next to white-tipped
black bill. Black cap; chestnut from
neck to blackish-brown underparts,
fading to paler brown flanks. Wings
brown above, white below. Winter
adult: crown, nape and back dark
brown. Wings brown. Rest of plumage
visible as it swims is buff.
DISTRIBUTION Widespread and
mainly resident in British Isles and
France, eastwards to S Baltic. Some
winter movement S and W to larger lakes and the coast by N and E birds.
HABITS AND HABITAT Breeds on inland waters with vegetated margins, even small
ponds or ditches. Secretive, often discovered only by its call. When disturbed, will dive
or scuttle across water rather than fly. Floating nest is attached to vegetation that touches
the water. Call: loud, whinnying.

Slavonian Grebe

■ *Podiceps auritus* L 31–38cm
WS 46–55cm

DESCRIPTION Red-eyed waterbird.
Summer adult: unmistakable head
pattern and colours, reddish neck
and body, black wings with small
white shoulder and white speculum.
Winter adult: crown down to eye
level, hind-neck and back black;
wings as summer. Rest of face,
neck and underparts white, sharply
contrasting with the black.

DISTRIBUTION Breeds in Iceland, around the Baltic and N Norway, plus a few in
N Scotland. Winters around W European coasts.
HABITS AND HABITAT Most breed on shallow lakes with plenty of vegetation. In
courtship, pair face each other and rear up in so-called 'penguin dance'. Nest is built of
water weed attached to aquatic vegetation and is often floating. Grey downy chicks have
black- and white-striped heads; they often ride on swimming parents' backs. Eats mostly
insects; also small fish.

Black-necked Grebe

■ *Podiceps nigricollis* L 28–34cm
WS 56–60cm

DESCRIPTION Has a steep forehead
and uptilted bill. Summer adult:
unmistakable head pattern and
colours, especially tuft of golden
feathers behind red eye. Neck and
upperparts dull black. Flanks chestnut,
belly white. Winter adult: crown down
to well below eye level, hind-neck and
back black; wings as summer, black
with white secondaries. Face, neck
and underparts white, contrasting
less than Slavonian (above).

DISTRIBUTION Breeds in scattered populations in British Isles, S Sweden and rest of
region. Winters around W European coasts.
HABITS AND HABITAT Prefers small, shallow waters with rich vegetation and a good
supply of insects and molluscs. Range insecure as land floods or dries out and populations
move. Usually breeds in colonies of up to 200 or more. Nest is a heap of water weed in
shallow water.

Great Crested Grebe

■ *Podiceps cristatus* L 46–51cm
WS 59–73cm

DESCRIPTION Easily spotted on open water, when neck is held erect. Summer adult: unmistakable, with thick V-shaped crest, and chestnut and black tippets (long feathers) surrounding white face. Long, dagger-like pinkish bill. Long neck, white before and blackish grey behind, edged with rufous. Flanks dark, tinged with rufous, rest of underparts silky white; back and wings brownish black, the latter with a white speculum and leading edge. Winter adult: loses tippets and crest; cap and nape black; head, face, foreneck and underparts white.

DISTRIBUTION Widespread breeder from British Isles (not N Scotland) to the Baltic. Winters on large lakes, reservoirs and the coast. In 19th-century Britain, almost became extinct; has since made a remarkable recovery.

HABITS AND HABITAT Breeds on large lakes, reservoirs and gravel pits. Nest is a floating heap of vegetation at the water's edge. Eats mostly fish.

Red-necked Grebe

■ *Podiceps grisegena* L 40–46cm
WS 77–85cm

DESCRIPTION Summer adult: black cap, white cheeks and chin; dark bill with yellow base; reddish-brown neck and breast, dark brown back and wings, the latter with a white speculum and leading edge. Winter adult: similar to winter Great Crested (above) but grubbier; note shorter neck and yellow-based bill.

DISTRIBUTION Restricted breeding in suitable habitats in E of region. Winters off North Sea and S Baltic coasts from S Norway to British Isles (especially the E coast).

HABITS AND HABITAT Prefers small, shallow lowland waters, often in association with colonies of gulls and other waterbirds. Nest is often further in lakeside reeds or sedges than other grebes would go. Eats adult and larval insects; also fish, especially in winter. Usually silent in winter, but when breeding has a loud wailing that sometimes turns into a scream.

Fulmar ■ *Fulmarus glacialis*
L 43–52cm WS 101–117cm

DESCRIPTION Gull-like with a white head and body, and grey upperwings, but it is a petrel – note the short yellowish bill with pronounced tubular nostrils, grey tail, and lack of black wing-tips. DISTRIBUTION Breeds in colonies on coastal cliffs from the English Channel N to Iceland and, in a few cases, E to Norway. Huge population growth in past 200 years as the amount of offal discarded from fishing fleets has increased. HABITS AND HABITAT Comes to land only to breed. Winters wholly at sea, to about 45°N. Flight distinctive, on stiff wingbeats followed by a glide on straight wings, usually low over sea. Pairs for life. 1 egg laid on bare rock. Chick flies after *c.* 6 weeks; lives at sea for 4–5 years, but will not breed successfully until at least 6 years. Mates cackle to each other on nesting ledge.

Manx Shearwater
■ *Puffinus puffinus*
L 30–35cm WS 71–83cm

DESCRIPTION A bit smaller than a Herring Gull (p. 77). Uniformly sooty black above and white below; white underwing edged black. Dark on head extends to just below eye. Sexes and ages similar. DISTRIBUTION A few small colonies in Iceland and Brittany; thousands in Faeroes, and *c.* 250,000 pairs in colonies around W and N British Isles. All birds migrate SW across the Atlantic to overwinter off the coast of South America; one fledgling is known to have reached Brazil in 17 days. HABITS AND HABITAT Glides effortlessly, or on stiff wingbeats, low over the sea, even in rough weather, rising and falling, looking black and then white. Nests in burrows on flat or sloping land next to the sea. Dives for fish and squid near the surface, and for offal from fishing boats. Comes to nest only at night to avoid predatory large gulls.

Leach's Petrel

▪ *Oceanodroma leucorhoa*
L 18–21cm WS 43–48cm

DESCRIPTION Larger and longer-winged than Storm Petrel (below). Black head, body and wings above and below, except for white rump and long, pale band across upperwing coverts. Forked tail, but fork not visible at all angles.

DISTRIBUTION A few colonies on Scottish islands, and maybe Ireland, Iceland, Faeroes and Norway. 10,000–100,000 in Scotland, but millions in Newfoundland. Winters at sea, S to South America and South Africa.

HABITS AND HABITAT Comes to land at night to breed, so very hard to census. Sometimes wrecked by being driven ashore in a gale. Feeds from the sea's surface on tiny crustaceans and other plankton, often pattering its feet on the surface as it does so. 1 egg is laid in a burrow or other hole. Incubation lasts 6 weeks, then chick fledges in 9–10 weeks, reaching maturity in *c.* 5 years.

Storm Petrel ▪ *Hydrobates pelagicus*

L 16cm WS 37–41cm

DESCRIPTION Tiny; wingspan about same as Song Thrush's (p. 116). Plumage sooty black, relieved only by broad white rump patch and white underwing bar. Looks very like a seagoing House Martin (p. 102).

DISTRIBUTION Breeds colonially in burrows on wild, undisturbed islands and islets in Iceland, but especially W Ireland, NW Scotland and the Faeroes. Winters in E Atlantic, as far S as Cape of Good Hope.

HABITS AND HABITAT Pelagic, over open sea, mostly offshore. Weak-looking, fluttering flight, just above the waves, feeding on small surface fish and plankton. Able to live well out to sea, but bad storms sometimes cause onshore wrecks. Will follow ships. In autumn especially, passage birds seen off W headlands. Breeders arrive at night to avoid predators. Fledglings find their own way to the sea. Purrs and grunts in burrow.

Gannet ▪ *Morus bassanus*
L 85–97cm WS 170–192cm

DESCRIPTION The region's largest seabird. Adult: mostly pure white but with black wing-tips, and yellowish-orange head in breeding season. Long, pointed wings; wedge-shaped tail; bluish-grey dagger-like bill, outlined in black. Juvenile: entirely grey-brown, covered with fine white speckles; amount of white increases, starting with head and body, until adult plumage reached after 4 years.
DISTRIBUTION Breeds in mostly remote cliff-top colonies in Iceland, Faeroes, Norway and Brittany, but mostly W and N British Isles. British and Irish population of *c.* 200,000 pairs is 70% of the world's total (most of remainder in E Canada). In winter, wanders as far as W Africa.
HABITS AND HABITAT Pelagic seabird of the N Atlantic continental shelf. Fishes spectacularly by diving steeply from 10–40m above sea, often in flocks.

Cormorant
▪ *Phalacrocorax carbo*
L 77–94cm WS 121–149cm

DESCRIPTION Care is needed to distinguish it from Shag (p. 21). Adult: all black, body glossed blue or green, wings glossed bronze. Long bill with yellow skin at base surrounded by patch of white. In spring has white thigh patch, and summer birds have some or much white on crown and nape. Juvenile: brown with grubby-white underparts; sometimes misidentified as a penguin.
DISTRIBUTION Breeds mostly in Iceland, British Isles and N Norway, and in scattered colonies from France to Poland. British birds disperse locally; more N birds migrate S.
HABITS AND HABITAT Nest of sticks and seaweed built in cliff colonies and (especially on Continent) in trees by lakes or the coast. Dives for fish, especially flatfish, with a leap out of the water (in contrast, divers just slip under head-first). Roosts on rocks, jetties, trees and sandbanks, with wings spread to dry. Flies strongly with neck outstretched.

Shag ▪ *Phalacrocorax aristotelis*
L 68–78cm WS 95–110cm

DESCRIPTION Smaller and slimmer than Cormorant (p. 20). Adult: all black save for greenish sheen and yellow at base of bill. Crest on forehead. Juvenile: brown, darker on back and wings, and without white underparts of Cormorant. Flight is with quicker wingbeats than Cormorant, but this is not easy to identify at a distance.
DISTRIBUTION Resident, with some local coastal dispersal after breeding, in Iceland, British Isles, NW France and Norway.
HABITS AND HABITAT Like Cormorant, swims low with neck erect. Breeds colonially, building a nest of twigs and seaweed on a sheltered ledge. Feeds on fish, diving with a leap forward; dive lasts c. 40 seconds. Usually hunts alone, but will fish in a flock where there is a shoal. Roosts singly or in small groups on rocks, often with wings spread.

Bittern ▪ *Botaurus stellaris*
L 69–81cm WS 100–130cm

DESCRIPTION Stocky brown heron with golden-brown neck, and black cap and moustache. Body and wings patterned with dark blotches and streaks, which splendidly camouflage it. In flight, the broad wings are 2-toned – flight feathers are dark brown and coverts buff. Sexes and ages similar.
DISTRIBUTION Scattered populations in suitable habitats from England to the Baltic. After breeding, many birds disperse to winter in W and S of breeding range.
HABITS AND HABITAT Secretive; found particularly in *Phragmites* reedbeds. Moves stealthily through reeds; when disturbed, freezes and points its yellow bill skywards, resembling reed stems. In a cold, prolonged winter, many may die. Hunted to extinction in British Isles by 1900 for its feathers, which were used in the clothing trade; a small number have bred here again since the 1950s. Male's spring call: peculiar, deep, repeated 'booming' note, which can carry 5km.

Little Egret ■ *Egretta garzetta*
L 55–65cm WS 88–106cm

DESCRIPTION Elegant all-white heron with pointed black bill, and black legs with contrasting yellow feet. Long neck is curled back when resting and flying, and feet project beyond tail in flight. Summer adult: has thin plumes on breast and back, and 2 long nape feathers. Immature: lacks plumes and toes less bright.
DISTRIBUTION Mostly a central and S European breeder. Numbers decimated by plumage trade in the 19th century. An unusual autumn influx occurred in Britain in 1989, with many birds staying; now >160 breeding pairs here, and c. 1,600 winter on sheltered estuaries. Also increasing in Brittany and the Netherlands.
HABITS AND HABITAT Colonial breeder, often with Grey Herons (below). Builds stick platform nests in trees or bushes, by marshes, lakes and lagoons. Eats small fish, plus crustaceans on the coast. Usually feeds singly, wading, sometimes energetically chasing prey.

Grey Heron ■ *Ardea cinerea*
L 84–102cm (neck extended)
WS 155–175cm

DESCRIPTION Adult: very long legs, long neck and dagger-like yellow bill. Back and coverts blue-grey, flight feathers dull black. Head white with black stripe over and behind eye, ending in long nape plumes. Body greyish white with black and white stripes down centre. Neck retracted in flight, with legs trailing beyond tail; wingbeats deep and slow. Juvenile: much greyer, lacking black and white contrast on head; bill greyish horn.
DISTRIBUTION Widespread resident in British Isles and France eastwards to the Baltic and Norway coast, but E birds migrate W and S.
HABITS AND HABITAT Usually solitary, but a colonial nester in tall trees ('heronries'). Eats mostly fish and eels from lakesides, rivers and estuaries. Call, often heard: loud, harsh *frank*.

White Stork

■ *Ciconia ciconia*
L 95–110cm WS 180–218cm

DESCRIPTION Huge, unmistakable bird. Adult: all-white body and wing coverts; black flight feathers. Red legs and long, pointed red bill. Juvenile: bill has dark tip.
DISTRIBUTION Declining. Scattered in E France and Low Countries, more common eastwards to stronghold on S side of the Baltic. Vagrant to British Isles. Much-loved bird of legend across Europe.
HABITS AND HABITAT Feeds in damp pastureland, marshes and riversides, on a wide variety of insects and animals. Breeds in solitary pairs or colonially, at a large stick nest used each year, in trees or on rooftops, towers and pylons. Summer visitor, wintering in tropical Africa. Large flocks gather at Mediterranean's narrows at Gibraltar and Bosporus, circling up on thermals before heading N or S; large flocks often also seen at plentiful food sources such as an insect swarm. Voice rarely heard, but loud bill-clappering used in pair's greeting ceremony.

Black Stork ■ *Ciconia nigra*
L 90–105cm WS 173–205cm

DESCRIPTION Adult: large, with black head, neck, breast, back and wings, and metallic green and violet gloss on neck and back. Rest of underparts and axiliaries white. Pointed red bill and red legs. Juvenile: dull black and white, and grey-green bill and legs.
DISTRIBUTION Summer visitor to E of the region; a few in Germany, and more as you go E. None breeds in British Isles or Scandinavia, where they are rare visitors. Winters in S Africa.
HABITS AND HABITAT Unlike White Stork (above), is a bird of undisturbed mixed forests, with streams, marshes, pools and swamps. Avoids human contact. Nests singly, high in crown of tree. Eats mainly fish and other aquatic creatures, caught by stalking. Voice and bill-clappering seldom heard.

Mute Swan ▪ *Cygnus olor*
L 140–160cm WS 200–240cm

DESCRIPTION One of region's largest birds (see also other swans). Adult: all white with black webbed feet and orange-red bill with a black base; knob at bill base is largest on the male ('cob'), which is a larger bird than the female ('pen'), as is true in the rest of the wildfowl. Long neck, pointed tail. Juvenile: grey-brown, bill dark grey; white plumage in following year.
DISTRIBUTION Widespread resident, especially in the British Isles; less common elsewhere. E birds migrate to moulting areas in W Baltic.
HABITS AND HABITAT Most W birds are probably feral; many introduced from the 16th and 17th centuries onwards on lakes, slow-flowing rivers and canals with shallow water for feeding. Now even in towns. Eats aquatic vegetation by dipping head or even up-ending. Builds huge nest on ground; male very protective and aggressive, with distinctive arched wings and hissing. In flight, wings make diagnostic loud, throbbing sound.

Whooper Swan
▪ *Cygnus cygnus*
L 140–160cm WS 205–235cm

DESCRIPTION Slightly smaller than Mute (above), and head usually held erect, not with Mute's curve. Adult: all-white plumage, head sometimes stained brown. Basal half of wedge-shaped bill yellow, finishing in a point, the rest black. Legs black. Juvenile: brownish-grey with off-white and pink bill.
DISTRIBUTION Breeds in Iceland and Fennoscandia. The latter populations migrate to winter on North Sea coasts; some Icelandic birds are resident, but c. 75% winter in Britain (mostly Scotland) and Ireland.
HABITS AND HABITAT Breeds in boggy tundra and upland lakes. Winters on lowland farmland. Pairs breed singly, family staying together in winter flocks. All flight feathers moulted together, so flightless for 5–6 weeks after breeding. Highly vocal, with bugling calls. Insignificant wing noise, unlike Mute.

Bewick's Swan

■ *Cygnus columbianus*
L 115–127cm WS 170–195cm

DESCRIPTION Adult: all-white plumage relieved only by black legs and yellow and black bill; yellow on bill generally a square or rounded patch, variable in shape, enabling observers to recognise individuals. Juvenile: grey, with pinkish-grey bill. Whooper (p. 24) and Bewick's both have square-ended tail.

DISTRIBUTION Winter visitor from Siberian breeding haunts, favouring areas in Britain, Ireland, and the Netherlands. One-third of population winters in the British Isles, especially at well-known reserves such as the Wildfowl Trust, Slimbridge, by the River Severn.

HABITS AND HABITAT Winters on low-lying wet pastures, saltmarshes and lakes. Feeds on aquatic plants, and grazes grass and waste potatoes and carrots. Families keep together for the winter. Adults pair for life.

White-fronted Goose

■ *Anser albifrons*
L 64–78cm WS 130–160cm

DESCRIPTION First of 4 so-called 'grey geese'. Adult: grey-brown body suddenly white under tail, belly with variable amount of black barring. Noticeable white blaze at base of pinkish bill. Has a narrow white band across base of tail, as do other 'greys'. Juvenile: lacks white blaze and belly marks.

DISTRIBUTION Winter visitor from Arctic Russia, mainly to England, Wales, Belgium and the Netherlands; and from Greenland (with predominantly orange bill) to Ireland and W Scotland.

HABITS AND HABITAT Winters in large flocks on low-lying wet grasslands, including farmland, but roosts on lakes, estuaries and floodwaters. Grazes and eats fallen grain and vegetable crops. Gregarious, sometimes in flocks of thousands, which fly in cackling, V-shaped formations to and from roosts.

Bean Goose ▪ *Anser fabilis*
L 69–88cm WS 140–174cm

DESCRIPTION Of the grey geese, the most difficult to separate on the ground from Pink-footed (below). Sooty-brown back, just as dark as flanks, and dark head. Orange and black bill, and orange legs. White under tail; narrow white edges to wing feathers show at close range. Upperwing almost uniformly dark. DISTRIBUTION In our region, breeds mostly in central Sweden and Arctic Finland. All populations winter mainly along S North Sea coasts, with a few in Britain.
HABITS AND HABITAT S birds are unusual for breeding in dense coniferous forest or birch scrub near water; N birds breed in characteristic low, wet tundra sites. Flightless for *c.* 1 month after breeding. Winter flocks graze on fields, and roost on lakes or flooded land near feeding ground.

Pink-footed Goose
▪ *Anser brachyrhynchus*
L 64–76cm WS 137–161cm

DESCRIPTION Basically grey-brown like other grey geese, but with a dark brown head and upper neck, paler below. Upperparts covered with white bars of feather edges, and has a frosty appearance; in flight, forewing is a striking blue-grey. Belly finely barred darker. Bill black with a variable pink band. Pink legs diagnostic (duller on juveniles). DISTRIBUTION In our region, a summer breeding visitor only in Iceland and Arctic Spitzbergen.
The former winter in Scotland and England, the latter in Denmark, Germany and the Netherlands.
HABITS AND HABITAT In Iceland, breeds in inaccessible river gorges to avoid predators, and in Spitzbergen on flat ground, grassy slopes and rocky outcrops. Flightless for 3–4 weeks after breeding. Wintering birds flock to estuaries, lakes and floodwaters to roost, and fields to feed on vegetable matter.

Greylag Goose

■ *Anser anser*
L 74–84cm WS 149–168cm

DESCRIPTION A stocky goose, with a thick neck and large orange bill. Back dark grey-brown, paler below, with narrow, darker stripes on flanks and down neck. Forewing and underwing coverts noticeably pale grey, very clear in flight. Wide white band at base and tip of tail. Legs pink.
DISTRIBUTION Originally bred over most of Europe, but hunting and drainage have reduced wild populations to Scotland, Poland, around the Baltic and coastal Norway. Reintroduced to England and the Low Countries.

HABITS AND HABITAT Feral British birds breed near freshwater lakes or on their islands, and feed on local grassland. Wild populations nest in marshes and wet tundra sites. Icelandic birds migrate to Britain; Scandinavian and E birds migrate SW as far as Spain. British birds disperse to safe British moulting areas, and are flightless for about a month. Ancestor of farmyard goose.

Canada Goose

■ *Branta canadensis*
L 90–100cm WS 160–175cm

DESCRIPTION Large brown goose, with black head and neck relieved by white blaze from behind eye to meet under chin. Wings dark above and below. White under base of tail. Bill and legs black.
DISTRIBUTION Native of North America; introduced to Britain in the 17th century, now widespread. Also feral in

the Netherlands, Belgium and Scandinavia. Mostly resident, but E birds migrate SW.
HABITS AND HABITAT Lowland lakes, even in city parks, where numbers can become a pest, fouling lakesides. Gander defends breeding territory aggressively, but species is gregarious outside breeding season, when flocks of hundreds or more form. Signals its approach in flight with deep, loud, trumpeting calls.

Barnacle Goose

▪ *Branta leucopsis*
L 58–70cm WS 120–142cm

DESCRIPTION Distinctive, with creamy-white face, and black cap, neck and breast. Rest of underparts white, with grey markings on flanks. Small black bill. Back and wing coverts blue-grey, striped black and white. Tail black. Black/white/grey pattern noticeable in flight. DISTRIBUTION Arctic breeder on islands and inaccessible cliffs. 3 distinct wild populations winter in our region: from Greenland to Ireland and NW Scotland; from Spitzbergen to Solway Firth; and from Siberia, via Baltic, to main area in the Netherlands.

HABITS AND HABITAT Feeds on coastal farmlands, which has put birds into conflict with farmers. Conservation efforts at present are helping both birds and farmers. Escapees from wildfowl collections now nest in Britain.

Brent Goose

▪ *Branta bernicla*
L 55–62cm WS 105–117cm

DESCRIPTION Small goose, the size of a Mallard (p. 30) but with a longer neck. At a distance, looks all black with a large white patch under tail. Closer up, small white broken neckband and paler belly are visible. 2 subspecies identifiable: 'dark-bellied', with belly almost as dark as back, but mottled; and 'light-bellied', with pale, mottled grey-brown belly. DISTRIBUTION Winter visitor from Arctic, breeding further N than any other goose: dark-bellied migrates from Siberian tundra to Netherlands, W France and England; and light-bellied from Greenland and Canadian Arctic to Ireland.

HABITS AND HABITAT Winter feeding grounds are along sea coasts and estuaries, where birds feed over mudflats, pulling up underwater plants, notably eel-grasses *Zostera*. In recent years, they have moved to graze on grassland. Will feed at night as well as day, because its feeding rhythm is controlled by tides.

Shelduck ■ *Tadorna tadorna*
L 55–65cm WS 100–120cm

DESCRIPTION Large, distinctive duck.
White with 'black' head (bottle-green
in good light), chestnut breast-band
and black flight feathers. Pink legs
and red bill, the male's larger than
the female's and with a noticeable
knob at base.

DISTRIBUTION Widespread around
region's coasts, less common in the Baltic and none in Iceland.

HABITS AND HABITAT In breeding season, occurs on sheltered coasts and estuaries.
Eats many species of invertebrate, especially the small snail *Hydrobia*, sifted from mudflats
by sweeping bill action. Breeding needs suitable nest-hole, e.g. in a sand-dune, riverbank
or Rabbit burrow. After breeding, migrates to traditional moulting areas, especially
Heligoland Bight and some British estuaries. Call: noisy *ak-ak-ak-ak* in breeding season.

Pintail ■ *Anas acuta*
L 51–62cm WS 78–90cm

DESCRIPTION First of the 'dabbling
ducks', so called for their habit of
up-ending, tail in the air, for food.
Breeding male: unmistakable, with
2 long (10cm), central tail feathers.
Head chocolate-brown with white
stripe reaching down to white breast.
Finely barred grey body, cream and
black under tail. Blackish-green
speculum with orange border at front
and broad white at rear. Female: light,
speckled brown, with a grey bill and
noticeably pointed tail; speculum less
noticeable than drake's but still with
clear white rear border.

DISTRIBUTION Breeds mainly in
Iceland, E Europe and Fennoscandia
eastwards; a few breed W to Britain.

TOP: *male;* ABOVE: *female*

Summer visitor, with many wintering
in British Isles and the Netherlands. Russian birds migrate as far as W Africa.

HABITS AND HABITAT In summer, inhabits wetlands with shallow water; winters on
sheltered coasts, estuaries and floodlands nearby. Drakes migrate to moulting areas as soon
as clutch is complete, e.g. hundreds to the Netherlands.

TOP: *male*; ABOVE: *female*

Mallard ■ *Anas platyrhynchos*
L 50–60cm WS 81–95cm

DESCRIPTION Male: dark green head, yellow bill, white neck-ring, purplish breast, grey belly and flanks, and black rear end with central tail feathers distinctively curled upwards. Orange legs. Female: camouflaged brown with darker markings; orange bill marked with black. Both have a dark blue speculum bordered with clear black/white borders. Both moult after breeding, when drake resembles duck with yellowish bill, for *c.* 4 weeks.

DISTRIBUTION Widespread and common throughout the region. Largely resident but retreats from ice.

HABITS AND HABITAT Ancestor of the farmyard duck, having been domesticated for centuries. Very adaptable, nesting in ground cover by lakes, ponds, canals and streams, from moorland to the coast and even in towns. It is the duck that quacks; drake has a rasping call and low whistle.

Gadwall ■ *Anas strepera*
L 46–56cm WS 78–90cm

DESCRIPTION Male: brown head; grey body, covered with tiny vermiculations; black tail end; greyish-black bill; long grey-brown scapulars. Has a duck-like moult plumage ('eclipse'). Female: resembles female Mallard (above), but with greyish head. Both have a white speculum, which often shows as they swim but is seen especially in flight.

DISTRIBUTION Has spread from Asia, in scattered populations as far as the British Isles and N to Iceland; occasional in S Scandinavia. Mostly resident in the W; E and Icelandic birds migrate to the S and W.

HABITS AND HABITAT Breeds on lowland lakes or slow-moving rivers. Nests on ground in dense vegetation. Usually silent.

TOP: *male*; ABOVE: *female*

Shoveler ■ *Anas clypeata*
L 44–52cm WS 73–82cm

DESCRIPTION Male: when seen swimming, has a dark green head and white–orange–white–black plumage from breast backwards, with long black and white feathers lying on dark brown back. In flight, shows striking pale blue forewing, blackish wing-tips and green speculum with broad white border in front. Female: speckled brown, with greyish forewing and paler speculum. Both have a noticeably long, broad bill.
DISTRIBUTION Scattered summer breeding populations in suitable habitats in Iceland, British Isles and France, eastwards to the Baltic and on across Asia. All birds migrate S or W, British birds to S Europe; in turn, Britain, Ireland and the Netherlands are major wintering areas for E birds.
HABITS AND HABITAT Marshes and shallow, small lakes. Feeds by reaching its huge bill forward, filtering tiny food through it as it paddles along.

TOP: *male*; ABOVE: *female*

Wigeon ■ *Anas penelope*
L 42–50cm WS 71–85cm

DESCRIPTION Breeding male: unmistakable. Head and neck chestnut with striking yellow forehead, pink breast, grey body and black tail end. In flight, shows green speculum and large white patch on forewing. Female: rufous-brown body, darker mottled back and rounded head. Both show a noticeable white belly in flight, and have a short grey bill with a black tip.
DISTRIBUTION Breeds in Iceland, Scotland and Scandinavia eastwards. British and Icelandic birds are resident or make only local movements; E birds winter in W and SW Europe.
HABITS AND HABITAT Nests solitarily by Arctic or sub-Arctic lakes, often by wooded areas, always where there is good cover. Winters gregariously, often in flocks of hundreds, on lakes, estuaries and deltas, to feed on extensive mudflats or sand bars, or graze on neighbouring grassland. Drake whistles *whee-oo*, duck purrs.

TOP: *male*; ABOVE: *female*

TOP: *male;* ABOVE: *female*

Teal ■ *Anas crecca* L 34–38cm WS 53–59cm

DESCRIPTION Noticeably small duck. Male: distinctive head pattern of dark green eye-patch outlined in gold on a chestnut background. Cream breast, speckled brown. Grey body with a white stripe dividing back from flanks, visible when swimming or waddling. Undertail yellow, bordered with black. Female: mottled brown. Both have delicate greyish bill, and in flight show a green speculum bordered with white.

DISTRIBUTION Widely distributed, breeding in Iceland, British Isles and France, eastwards across Asia. W populations winter in S Europe, N and E populations mainly in the Netherlands and British Isles.

HABITS AND HABITAT Nests very close to water, in marshes and wet upland moorland. Winters in small flocks on lakes, reservoirs and coastal lagoons. Drake calls melodious *prip-prip*, duck quacks.

TOP: *male;* ABOVE: *female*

Garganey ■ *Anas querquedula* L 37–41cm WS 59–67cm

DESCRIPTION Male: not brightly coloured but distinctly patterned. Dark brown head with broad white stripes over each eye that meet behind neck; grey flanks, long black and white scapulars droop over brown back. In flight, forewing is pale blue-grey, and speculum is dark green with broad white borders. Female: separated from similar female Teal (above) by longer grey bill, striped head pattern (especially contrasting pale supercilium over dark eye-stripe) and greyer wing.

DISTRIBUTION Summer visitor to W Europe (including England), around the Baltic and E across Asia. Long-distance migrant, wintering in Africa S of the Sahara.

HABITS AND HABITAT In summer and winter, found in water meadows, and marshy and shallow freshwater pools with emergent vegetation. Not often seen on coasts or estuaries. Feeds as it swims, on plants and aquatic creatures.

Goldeneye ■ *Bucephala clangula*
L 40–48cm WS 62–67cm

DESCRIPTION First of the 'diving ducks'. Male: domed head glossy green, with white spot at base of grey bill and yellow eye. White breast, belly, flanks, speculum and wing coverts; black back and tail end. 'Eclipse' male as female, some with white loral spot. Female: brown head with pale yellow eye, and white collar; rest of plumage mottled grey; in flight, shows less white on forewing than male.
DISTRIBUTION Summer visitor to Fennoscandia, Russia and across N Asia. In our region, most winter in the Baltic W to British Isles.
HABITS AND HABITAT Breeds mostly in tall forest close to lakes or rivers with open water. Nests in a natural tree-hole, and readily uses nestboxes up to 5m above ground. Ducklings tumble to ground after hatching, and are led to the safety of water, where they can feed themselves. Dives for aquatic animal food. Winters on lakes, reservoirs and sheltered coastal waters.

TOP: *male;* ABOVE: *female*

Pochard ■ *Aythya ferina*
L 42–49cm WS 72–82cm

DESCRIPTION Male: distinctive chestnut-red head, light grey back, and black breast and tail end. Female: head and breast brown, back tinged grey, rest of underparts white with grey markings. Both have a blackish bill with a broad, central blue-grey band; in flight, they show a long greyish-white wing bar.
DISTRIBUTION Across the British Isles to the Baltic and into Asia. British population is augmented in winter by migrants that come from the E.
HABITS AND HABITAT In breeding season, found on lakes, large ponds and slow-flowing rivers with cover. Nests in or very close to water on a mound of vegetation built by female. In winter, frequents open waters of lakes, reservoirs and estuaries, but rarely on open sea. Dives mainly for aquatic plants.

TOP: *male;* ABOVE: *female*

TOP: *male*; ABOVE: *female*

Scaup ▪ *Aythya marila*
L 42–51cm WS 71–80cm

DESCRIPTION Male: as it swims, looks black at the front and grey-backed, with white flanks and a sloping black tail end. Head has a green sheen, eye yellow and bill grey with a black tip. Female: dark brown head and breast, mottled brown-grey back and flanks, and a broad white band at base of bill.
DISTRIBUTION Breeds in Iceland, Fennoscandia, around the Baltic and into Siberia. In winter, migrates to S Baltic coasts and W to British Isles.
HABITS AND HABITAT Very gregarious for much of the year. Breeds near water in tundra, even in association with gull or tern colonies. Winters mostly on the coast, sometimes in huge flocks of hundreds over a good food supply of molluscs a few metres deep, for which it dives.

TOP: *male*; ABOVE: *female*

Tufted Duck ▪ *Aythya fuligula*
L 40–47cm WS 65–72cm

DESCRIPTION Male: all black save for pure white flanks and golden eyes. Has a drooping crest at back of head. Female: mostly dark brown, with some white on sides and usually only a little white at base of bill (do not confuse with female Scaup, above; note small tuft, less white by bill and rounded head shape). Both adults have a long white wing bar on dark wings.
DISTRIBUTION In suitable habitats across the region, including Iceland. S birds are largely resident, but Icelandic and E birds migrate to milder S and W in Europe.
HABITS AND HABITAT Breeds on freshwater lakes and rivers with fringes of good cover. Nests on ground near water, even in gull and tern colonies. Winters on open but sheltered waters. Tolerant of human presence and found on reservoirs and city parks.

Eider ■ *Somateria mollissima*
L 60–70cm WS 95–105cm

DESCRIPTION Large duck. Adult male: white with black crown, flanks, belly and tail – an unusual white-above and black-below appearance. Sides of head are lime-green; large, wedge-shaped olive-coloured bill with distinctive wedge of white feathers pointing forward from base. In eclipse plumage, sooty brown with white lower breast and wing coverts. Adult female: brown with darker bars and mottling. Juvenile: drakes take 4 years to get adult plumage, so have a patchy black and white plumage.
DISTRIBUTION Wholly coastal, roughly N of 55°N, from N England to Arctic and then across region.
HABITS AND HABITAT Confined to marine

TOP: *male*; ABOVE: *female*

habitats along rocky coasts, where it dives for Blue Mussels *Mytilus edulis*, other shellfish and crabs. Highly gregarious at all times, and can be seen offshore for much of year. Courtship starts in winter, with males throwing heads back and cooing (!) *a-HOOO, a-HOOO*. Often nests colonially; after hatching, young form crèches with several female 'guardians' or 'aunties'. Has been protected for centuries in England, Iceland, Norway and elsewhere to maintain collection of nest down for sleeping bags and eiderdowns.

King Eider ■ *Somateria spectabilis* L 55–63cm WS 87–100cm

DESCRIPTION Male: pink bill with prominent orange knob outlined in black at base, pale blue head, black body with curled tertials looking like 2 small sails. Female: similar to female Eider (above) but more rufous and with dark bill.
DISTRIBUTION Breeds N of Arctic Circle on Spitzbergen and in Russia. Winters around the coasts of Fennoscandia, Iceland and, rarely, N Britain.
HABITS AND HABITAT Breeds on tundra pools, unlike Eider, often far from coast. Short summers of *c.* 60 days in breeding areas mean pools start to freeze early, so families often have a long walk to coast to survive.

LEFT: *male*; ABOVE: *female*

TOP: *male;* ABOVE: *female*

Common Scoter ■ *Melanitta nigra*
L 44–54cm WS 70–84cm

DESCRIPTION Male: glossy black all over except for yellow patch on bill. Female: dark brown, save for whitish sides of face. At a distance in some lights, both look black.
DISTRIBUTION A few breed in Iceland, Ireland and Scotland; breeds more widely in Fennoscandia and across Russia. Winters around British coasts, S Baltic and coasts westwards to Iberia.
HABITS AND HABITAT Nest well concealed in tundra, even well inland and away from water, or by moorland lochs or slow-flowing rivers. In winter, found mostly on shallow inshore marine waters where it dives for molluscs and crustaceans. Gregarious in winter, with many hundreds in a flock; often seen flying low over the sea in long lines.

TOP: *male;* ABOVE: *female*

Velvet Scoter ■ *Melanitta fusca*
L 51–58cm WS 90–99cm

DESCRIPTION Male: glossy black with white secondaries seen well in flight and sometimes as a white patch on side as bird swims. Bill orange with a black centre. Female: dark brown above, paler below, with white secondaries and indistinct white on side of head.
DISTRIBUTION Breeds across Fennoscandia, into Russia. Winters along Norwegian coast, S Baltic (many thousands moult off Denmark), E coast of Britain and English Channel coast.
HABITS AND HABITAT Breeds by inland lakes and pools, and by Baltic shores, all in wooded areas. In winter, is the least numerous of the sea ducks; may well be overlooked among movements of the much more numerous Common Scoter (above). Diet comprises mainly molluscs, but is more varied than that of its relative.

Long-tailed Duck ■ *Clangula hyemalis*
L 39–47cm (but see below) WS 65–82cm

DESCRIPTION Summer male: dark brown head, breast and back, head with white spot behind eye and back streaked with yellow-edged scapulars; wings all dark. Long, central tail feathers add 10–15cm to length. Summer female: paler version of summer male, with short tail. Winter male: white head and body, with brown eye-patch, dark brown wings and brown lower breast. Winter female: dark cap, white face with brown patch below eye, warm brown breast, rest of underparts white, back brown.
DISTRIBUTION Breeds on Arctic tundra of Iceland, Fennoscandia and, most commonly, eastwards across Siberia. Winters on N coasts, including British Isles. Most numerous duck in far N.
HABITS AND HABITAT Breeds on tundra pools, coastal islands and by ice-free mountain pools. Gregarious in winter, often in very large flocks, e.g. up to 10,000 in Moray Firth, but very few further S. Dives for molluscs and crustaceans.

TOP: *winter male*; ABOVE: *winter female*

Smew ■ *Mergus albellus* L 38–44cm WS 55–69cm

DESCRIPTION The first of 3 'sawbills', so called because their bills have a serrated edge. Male: appears predominantly white when swimming, but with black face patch; slight crest is white above, black below; narrow black inverted 'V' on sides of breast; black centre to mantle. In flight, appears pied, with black and white wings and body. Female: grey back, paler grey underparts; distinctly patterned head, with chestnut cap and nape sharply divided from white cheeks and throat.
DISTRIBUTION Breeds only in Arctic areas of region. Winters to SW in small numbers from Denmark to England.
HABITS AND HABITAT Nests in a hole in a mature tree (often a woodpecker hole) or nestbox, close to a lake or slow-flowing river. In winter, found mainly on freshwater ponds, streams and reservoirs. Often associates with Goldeneyes (p. 33). Expert diver for small fish, which it holds tight in its specialist bill.

TOP: *male*; ABOVE: *female*

TOP: *male;* ABOVE: *female*

Goosander ■ *Mergus merganser*
L 58–68cm WS 78–94cm

DESCRIPTION Largest sawbill. Male: looks long-bodied when swimming; black back; white body, tinged pink; black, glossed green head with steep forehead and mane-like crest. Bill broad-based then long and fine, red and hook-tipped. White secondaries and coverts show in flight. Female: chestnut head, sharply divided from white throat and white underparts. Rest of body grey. In flight, shows white secondaries, grey coverts and blackish primaries.
DISTRIBUTION Breeds in Scotland, then spreads to N England and Wales, with a few further S; also Iceland, Baltic, Scandinavia and Russia. British and Icelandic birds are largely resident. Others migrate W to winter around the Baltic, North Sea and Britain.
HABITS AND HABITAT Breeds in hilly country beside upland lakes and slow-flowing rivers near forests. Nests in a tree-hole, tree stump, rock crevice or nestbox. Sometimes winters in large flocks, even fishing communally, on large, open waters. Gets into trouble with commercial fisheries.

TOP: *male;* ABOVE: *female*

Red-breasted Merganser
■ *Mergus serrator* L 52–58cm WS 67–82cm

DESCRIPTION Male: striking shaggy crest from back of black (glossy green) head. White neck-ring, thin red bill; chestnut breast spotted with black, side of breast black with white spots, flanks and tail grey, back black, innerwing white, primaries dull black. Female: reddish-brown head, fading into grey body. Crest less noticeable than drake's.
DISTRIBUTION Widespread from N British Isles and Iceland, eastwards across Scandinavia. In winter, E birds migrate W; others disperse from inland breeding sites.
HABITS AND HABITAT Usually seen on the sea in winter, having bred beside a lake or slow-flowing river near woodland. Nest is on ground in good cover.

White-tailed Eagle

■ *Haliaeetus albicilla*
L 76–92cm WS 190–240cm

DESCRIPTION Very large; female is usually bigger than male, as in most raptors. In flight, wings are held almost level, with spread primary tips ('fingers'); this shape has led to the bird being called a 'flying plank'. Adult: mostly dark brown, head paler, coverts and back feathers tipped yellowish; large, deep, yellow bill; all-white wedge-shaped tail. Immature: more uniformly dark, including tail, which is not white until *c.* 5 years old; bill grey-black.

DISTRIBUTION Rare, with a small number of pairs in Iceland, Germany, Poland and Scandinavia eastwards. Widespread in British Isles in the 18th century but extinct by 1916 through persecution (Continental birds also suffered). In 1975, young Norwegian birds used to reintroduce species to Scotland; these now number *c.* 40 wild pairs.

HABITS AND HABITAT Mostly resident around rocky coasts and islands. Nests on crown of a mature tree or, less often, on a cliff ledge. Some eyries used for many years. Predator on fish, waterbirds and mammals; scavenger of carrion.

Osprey ■ *Pandion haliaetus*
L 52–60cm WS 152–167cm

DESCRIPTION In flight, wings are often seen as a shallow, gull-like 'M'. Adult: long-winged, with white head and underparts, dark brown upperparts and upperwing. Short, square-cut tail. Speckled breast-band, thick black band through yellow eye. Juvenile: distinguished by buff tips to mantle and wing coverts.

DISTRIBUTION Persecuted to extinction in much of Europe by early 20th century. Naturally returned to Scotland in 1955 after *c.* 40 years, now *c.* 150 pairs; only widespread eastwards from Scandinavia. Seen more widely in autumn, usually singly, while on migration to Africa S of Sahara.

HABITS AND HABITAT In all locations, found on lakes, rivers and coast wherever fish can be caught in talons after shallow dive; gets into trouble at fish farms. Builds a large stick nest on a treetop or specially erected pole, often using it year after year.

Golden Eagle
■ *Aquila chrysaetos*
L 80–93cm WS 190–225cm

DESCRIPTION Very large raptor. In flight, wing-tips are 'fingered' and wings have slightly curved rear edges; in head-on view, wings appear in shallow 'V'. Adult: uniform dark brown above, except for golden cowl around back of head and neck; wing coverts become bleached, resulting in paler upperwing panel. Underwing flight feathers and tail dark grey, barred darker, especially tips. Juvenile: lacks the gold; tail white with broad black tip; base of underwing primaries and secondaries white; some white lost each year for *c.* 5 years.

DISTRIBUTION Marked decreases since the 19th century through persecution. In suitable habitats in Scotland and Scandinavia eastwards. With protection, there are now >400 pairs in Scotland, the densest population in the region. Mostly resident, but far N birds disperse and immatures migrate further.

HABITS AND HABITAT A bird of mountains, moorlands and, in the E, forested areas. Preferred prey is grouse and hares, but also eats other mammals and medium-sized birds, plus carrion. Builds a huge nest on crag or tree, often reusing it.

Lesser Spotted Eagle
■ *Aquila pomarina*
L 55–65cm WS 143–168cm

DESCRIPTION Adult: dark brown body and flight feathers, paler coverts above and below, and white patch showing at base of primaries on upperwing. Juvenile: is spotted – tips of wing coverts form 2 narrow white bars.

DISTRIBUTION Has seen a marked contraction of range; now in E Germany eastwards around S Baltic. Summer visitor, wintering entirely in E Africa.

HABITS AND HABITAT Mainly remote lowland forests by marshy meadows. Hunts for small mammals by low, gliding flight, quartering the ground. Has declined through deforestation and persecution. Solitary or in pairs in breeding season, but large numbers can be seen crossing Bosporus in autumn.

Black Kite ■ *Milvus migrans*
L 48–58cm WS 130–155cm

DESCRIPTION Rather dingy, dark brown plumage relieved only by paler head and wing coverts, grey-brown tail, and paler bases to underside of primaries. Slight fork does not show when tail is spread. DISTRIBUTION Summer visitor, especially to S part of region, including from Germany eastwards around S side of Baltic into Russia. Winters in Africa S of Sahara. HABITS AND HABITAT Likes the immediate neighbourhood of rivers and other wetlands, and around human

settlements. Predator and scavenger on a wide range of prey and carrion, and will chase other birds to get them to drop food. Nests in a tree. Most W European birds migrate via the Straits of Gibraltar, where several thousands can be seen in a day in early autumn.

Red Kite ■ *Milvus milvus*
L 61–72cm WS 140–165cm

DESCRIPTION Long-winged, with a strikingly forked tail. Adult: underparts and back deep rufous brown, streaked darker; head pale greyish, streaked darker; tail rufous above, paler below. Upperwing has yellowish diagonal bar between brown coverts and dark flight feathers. Underwing dark grey-black except black tip, large white panel on rest of primaries, and rufous brown on coverts. Juvenile: white tips to upper greater coverts; underparts yellowish buff and less heavily streaked than adult, so looks paler. DISTRIBUTION Now virtually confined to Europe. Occurs across our region in small numbers to S Sweden. Has been much persecuted. HABITS AND HABITAT Open woodland by farmland, rough pasture and heath, where it hunts for live prey and carrion. Used to be common in cities, scavenging. Reduced in Britain to a few pairs in Wales by the mid-20th century. Reintroduced in 1989 and widely since then; now *c.* 1,000 breeding pairs. Mainly migratory in N, to SW as far as Mediterranean; others sedentary.

Marsh Harrier ■ *Circus aeruginosus* L 43–55cm WS 115–140cm

DESCRIPTION All 3 harriers are long-winged, long-tailed and long-legged. Adult male: head and breast buff, streaked darker; belly chestnut brown. Wings appear tricoloured: tips black, secondaries and some coverts grey, other coverts brown. Tail plain grey. At some angles can resemble male Hen Harrier (below). Adult female: dark brown with creamy-white crown and throat; wing coverts pale brown, especially leading edge. Juvenile: golden crown and throat.

DISTRIBUTION In suitable habitats from England (a few) to the Baltic eastwards. Summer visitor, favouring aquatic habitats; winters in Africa S of Sahara.

LEFT: *male;* ABOVE: *juvenile*

HABITS AND HABITAT Specialist raptor of reedbeds and marshes. Hunts by quartering slowly over reeds and dropping on a variety of prey. Nests on ground in rank vegetation. Male passes female food in flight during courtship, foot to foot, or dropped for her to catch, and also later for the chicks. Alarm *kek-ek-ek;* whistles during aerial courtship display.

Hen Harrier ■ *Circus cyaneus* L 45–55cm WS 97–118cm

DESCRIPTION Adult male: pale grey with white belly and uppertail coverts, black wing-tips. Adult female: upperparts brown with white rump; underparts whitish, streaked brown; tail grey with 5 bold, dark bars. Underwing coverts brown; flight feathers pale grey with clear black bars along length of wing.

Juvenile: as female but plumage more rusty with dark underwing secondaries.

DISTRIBUTION In suitable habitats in British Isles and across region. Migratory in N and NE to winter within Europe.

HABITS AND HABITAT Breeds in open taiga, moors and heaths. Winters in other open country, including fens, dunes and fields. Nests on ground. Male catches food for young and passes it to female in the air, calling *kek-kek-kek.*

Montagu's Harrier

■ *Circus pygargus*
L 39–50cm WS 96–116cm

DESCRIPTION Adult male: grey, similar to male Hen Harrier (p. 42) but has black band across upper secondaries, darker grey back and wing coverts, and much less noticeable pale rump; rufous streaks on flanks. Adult female: very similar to female Hen Harrier, but slimmer and narrower-winged, with heavily barred underwing coverts and often less noticeable white rump. Juvenile: like female but underparts chestnut.

DISTRIBUTION Summer visitor to suitable habitats in low numbers, scattered from S Britain to France and across region. Winters in Africa S of Sahara, especially in E.

HABITS AND HABITAT Found on open moors, rough ground, fens, heaths and young forestry plantations. Nests on ground in thick vegetation. Catches a wide variety of food, from large insects to small birds and mammals. Male provides most of food for young, but it is fed to them by female.

TOP: *male*; ABOVE: *female*

Rough-legged Buzzard

■ *Buteo lagopus* L 50–60cm WS 120–150cm

DESCRIPTION First of 5 raptors known as 'hawks'. Closely resembles Buzzard (p. 44), both species having variable plumage. Adult: generally has whitish head and underparts, dark belly, largely white underwing with black tips and carpal patch, and white tail with black terminal band.

DISTRIBUTION Summer visitor to Fennoscandia and N Russia. Winters in S Sweden and Denmark southwards; a few occur each year in E Britain.

HABITS AND HABITAT Nests mostly on rocky ledges or sheltered sites on ground in low-lying, treeless tundra. Numbers and distribution there vary in relation to abundance or scarcity of small mammalian prey, especially lemmings. Winters in open country. Often hovers, unlike its relative.

Buzzard ■ *Buteo buteo*
L 51–57cm WS 113–128cm

DESCRIPTION Upperparts dark brown, underparts white with very variable amount of brown streaks and bars, most noticeable across breast. Pale underwings with dark carpal patch. Glides with wings in noticeable 'V'.

DISTRIBUTION Found across region, but rarely in Eire and not in Iceland or N Fennoscandia. Population is increasing in England. Most W birds are resident; Scandinavian and E birds migrate to winter S and W as far as Mediterranean.

HABITS AND HABITAT Most often seen, especially in spring, as it spirals over its territory on farmland, forest clearings and moorland edges with mature trees. Heavily persecuted still in some places in belief it harms human interests, but its main food is small rodents. Often hunts by waiting to drop onto prey passing below its perch, but will walk a field for earthworms. Builds a bulky tree nest of sticks with a soft lining, decorated repeatedly with fresh leafy sprays. Calls freely throughout the year, a mewing *pee-oo, pee-oo*.

Honey Buzzard
■ *Pernis apivorus*
L 52–60cm WS 135–150cm

DESCRIPTION Distinctive flight silhouette, with a small head on a slender neck, slight bill and long tail. Male: grey-brown above with ash-grey head and yellow eye. Underparts heavily barred. Underwing typically white with a black carpal patch and trailing edge, and noticeably barred coverts and tail. Female: similar but darker.

DISTRIBUTION Summer visitor, mostly to E of region, with only a few in Norway and S England. Winters in wooded equatorial Africa.

HABITS AND HABITAT Secretive bird of mature, mostly deciduous woodland. Specialist feeder on bees and wasps, including their larvae and pupae, dug out of nest with strong talons. Walks and flies freely in thick woodland.

Sparrowhawk

■ *Accipiter nisus* L 28–38cm WS 55–70cm

DESCRIPTION In flight, shows short, rounded wings and long tail. Flies with rapid burst of wingbeats, then a glide, revealing barred underwing and tail. Male: grey above; underparts pale, narrowly barred orange. Female: up to 25% bigger than male. Brown above, barred brown below.

DISTRIBUTION Found throughout region except Iceland and far N. N Scandinavian birds migrate to winter within rest of range; others resident or disperse locally.

HABITS AND HABITAT Preys almost entirely on birds. Hunts by dashing from perch to perch in woods, farmland and gardens, snatching prey in flight or off its perch – including bird tables! Male takes small songbirds; female can kill a dove. Quite often seen being mobbed by Swallows or Starlings above hunting ground.

Goshawk ■ *Accipiter gentilis*
L 48–62cm WS 135–165cm

DESCRIPTION Grey-brown above; greyish white below, closely barred brown, but noticeably white under tail, which has 4 or 5 bars. Clearly patterned head – dark crown, white supercilium, dark ear coverts, white chin. Heavily barred underwing.

DISTRIBUTION Widespread resident from Germany eastwards; some dispersal from coldest N. A few hundred pairs in Britain, many believed to be released by falconers.

HABITS AND HABITAT Exclusively a forest bird. Has suffered badly from human persecution, e.g. gamekeepers. Nests in treetops with clear access, such as on edge of a glade. Even though it is large, it is adept at pursuing prey as big as a Capercaillie through trees.

Kestrel ■ *Falco tinnunculus*
L 32–35cm WS 71–80cm

DESCRIPTION The first of 5 'falcons'. Male: chestnut upperparts with blackish spots; dark-streaked buff underparts; bluish-grey head, rump and tail, the last with a black sub-terminal band and white tip. Female: reddish brown above with dark barring, paler below with dark streaks, and a barred tail. Pointed wings (cf. rounded tips of hawks).
DISTRIBUTION Throughout region except Iceland. Mostly resident, except in far E and N of region, whose birds migrate to within rest of region. In many places it is the commonest daytime raptor.

HABITS AND HABITAT Easily seen as it habitually hovers looking for prey, suddenly dropping onto it. Then moves on to hover again, over farmland, hill country, moorland or rocky coasts. Nests in old crows' nests, less commonly on cliff ledges, hollow trees, large nestboxes and ruined buildings.

Hobby ■ *Falco subbuteo*
L 30–36cm WS 82–92cm

DESCRIPTION Comparatively short tail and long, scimitar-shaped wings – can look like a large Swift (p. 96). Sexes similar. Dark slate-grey above, heavily streaked below; patterned head of dark cap, whitish cheeks and throat broken by dark moustachial stripes; diagnostic rufous thighs and undertail.
DISTRIBUTION Long-distance summer migrant to S Britain eastwards to around the Baltic. Thinly distributed in the W. Winters in S third of Africa in wet season, when insects are abundant.
HABITS AND HABITAT Prefers open lowland habitats, where it can pursue its almost exclusively aerial prey – insects and birds, especially where abundant, such as a bird roost or insects at a bush fire. Nests in old crows' nests or similar. Male provides most of food for young, especially 'open sky' birds such as swifts, swallows, martins, larks and pipits.

Peregrine ■ *Falco peregrinus*
L 36–48cm WS 95–110cm

DESCRIPTION Powerful-looking raptor with broad-based, pointed wings and tapering tail. Adult male ('tiercel'): dark grey-blue upperparts, including wing coverts and tail; buffy-white underparts, finely spotted and barred black. Head patterned with blue-grey moustache and cap to below eye, outlined by white cheeks and throat. Adult female ('falcon'): 15% larger than male, darker and more heavily barred. Juvenile: browner above and underparts streaked, not barred.
DISTRIBUTION Not Iceland; from British Isles eastwards in suitable habitat. Marked declines in the mid-20th century due to poisoning by organochlorines in the food chain. Has recovered well with protection in the British Isles.
HABITS AND HABITAT Chiefly open country with coastal or inland cliffs, including quarries, for a nesting ledge. Birds are main prey (especially seabirds and pigeons, the latter resulting in persecution by pigeon fanciers); hunts by circling high, spotting prey below, then diving at incredible speed for aerial kill.

Merlin ■ *Falco columbarius*
L 25–30cm WS 50–62cm

DESCRIPTION Male: about the size of a Mistle Thrush (p. 115); upperparts (including tail) slate-blue, underparts rusty, streaked with brown; tail has black terminal band. Female: larger, with dark brown upperparts and whitish underparts, streaked dark brown; tail is boldly striped brown and cream.
DISTRIBUTION Breeds in the British Isles, Iceland, Fennoscandia and thence across Asia. British birds are mostly resident, augmented in winter by Icelandic birds; Scandinavian birds winter in mainland Europe, SW from S Baltic.
HABITS AND HABITAT Breeds on hill country not dominated by trees. In winter, found on open country generally and low-lying coasts such as sand-dunes. Nests on ground in heather, or sometimes in old crows' nests. Mostly preys on small birds like pipits, larks and newly fledged moorland nesting waders. Hunts in level flight, perhaps only a metre off the ground, and catches prey in surprise attack.

Gyrfalcon ■ *Falco rusticolus*
L 50–60cm WS 130–160cm

DESCRIPTION Large, powerful raptor with very variable plumage. Typical S bird has upperparts blotched and barred overall with dark and light grey; underparts are whitish, heavily streaked blackish. N birds are paler and greyer. Very rarely, a pale form from Greenland, pure white, lightly spotted with black, appears in winter.
DISTRIBUTION Iceland and Arctic Europe. Disperses to coast or S to *c.* 60°N after breeding. Rare winter visitor to British coasts.
HABITS AND HABITAT Inland rocky valleys, coastal cliffs and mountains. Prey is mainly birds – seabirds on coast, and grouse family members especially inland – and more mammals than the Peregrine (p. 47), including voles and hares.

Willow Grouse ■ *Lagopus lagopus* L 37–42cm WS 55–66cm

DESCRIPTION Two main subspecies occur in region: Willow Grouse *L. l. lagopus* and Red Grouse *L. l. scoticus*. Breeding male Willow: all upperparts and head to chin a rich red-brown, barred darker; wings and underbody white; tail black. Breeding female Willow: as male but barring more obvious and chin white. Winter Willow: both sexes become all white except black tail. Male Red: never has white wings; entire plumage a rich red-brown, except for white underwing coverts and variable amount of white on underparts in winter. Female Red: paler than male. Both sexes of both subspecies have red wattles over eyes, but these more noticeable in males, especially high-ranking individuals.
DISTRIBUTION Willow resident across Scandinavia and into Asia. Red resident in W British Isles. Population cycles related to food supply.
HABITS AND HABITAT Willow breeds in tundra, moors, dwarf willow and birch; Red confined to heather moors. Both almost entirely vegetarian, Red feeding mostly on heather throughout the year. Both hunted; Red has long been husbanded on 'grouse moors'. Call often heard: *go-back go-back back back*.

LEFT: *male Red Grouse;* ABOVE: *male Willow Grouse*

Ptarmigan
■ *Lagopus mutus*
L 34–36cm WS 54–60cm

male summer

DESCRIPTION Breeding male: upperparts dark grey-brown and black, finely barred with white; greyer in autumn. Wings and underparts white. Winter male: white, feathered from nostrils to toes, except for black tail and lores. Summer female: golden-yellow markings in the brown. Winter female: white; no black lores.

DISTRIBUTION Iceland, Scotland and Fennoscandia in suitable habitats. Resident, even in January at –35°C, but retreats from deep snow.

HABITS AND HABITAT Prefers open rocky or stony ground, which may be near sea-level in Arctic, but in hilly or mountainous country further S, usually *c.* 1,000–1,200m. Often very approachable. Feeds on a wide range of plant food; in winter, digs in snow to find it. Reluctant to fly; crouches, wonderfully camouflaged. Gregarious for much of year, otherwise in pairs or families.

Hazel Grouse
■ *Bonasa bonasia*
L 35–37cm WS 48–54cm

DESCRIPTION Intricately patterned, partridge-sized bird. Male: greyish upperparts, black-bordered tail; rufous breast, whitish belly, both with black and rufous scallops; upperwing greyish with rich pattern of brown and black; black bib, bordered white, and small crest. Female: smaller crest; brown throat speckled with white.

DISTRIBUTION A few in Germany, Poland and Belgium, then around the Baltic, Fennoscandia and Russia in suitable habitat. Sedentary.

HABITS AND HABITAT Shy. Only found in extensive, undisturbed forest (usually mixed-species coniferous with a good shrub layer). Feeds on a wide variety of buds, seeds, leaves and fruits, in winter mostly in trees, and at other times on ground. Solitary or in pairs. Flushed birds fly with a distinctive whirring sound of wings. Call: high-pitched whistle, more like a warbler.

Capercaillie ▪ *Tetrao urogallus*
L 60–87cm WS 87–125cm

DESCRIPTION The largest grouse. Male: *c.* 40% larger than female. Blackish body and tail, brown wings; black is tipped white on shoulders and upper-tail coverts. Female: brown, barred and mottled buff, black and white; large size, rufous-orange breast patch and rounded tail separate it from greyhen (below).
DISTRIBUTION Scotland, Fennoscandia and around the Baltic. Became extinct in Britain in the 18th century, reintroduced in the 19th. Decreasing in many areas owing to deforestation, disturbance and shooting.
HABITS AND HABITAT Almost confined to coniferous forest with glades. In Oct–Apr eats shoots and buds of conifers picked off in the trees' crowns; rest of year feeds mostly on ground on a variety of vegetable foods, especially berries in autumn. Males display, raising and fanning turkey-like tail, but not so organised as blackcocks' lek. Well-camouflaged female nests on ground and tends chicks alone. Male has extraordinary 'song' lasting *c.* 5 seconds, made of 3 or 4 different sounds running into each other: quiet tapping, getting faster, followed by *pop* and ending with a grinding sound.

Black Grouse ▪ *Tetrao tetrix*
L 40–55cm WS 65–80cm

DESCRIPTION Male ('blackcock'): unmistakable. Glossy black with white wing bar and under-tail, lyre-shaped tail. Female ('greyhen'): warm brown above, paler below, speckled with black; forked tail; white wing bar distinguishes it from Willow and Red grouse (p. 48) and Capercaillie.
DISTRIBUTION In suitable habitats in British Isles (not Ireland) eastwards around the Baltic into Fennoscandia and Russia. Resident, but decreasing almost everywhere except Norway and Sweden through habitat destruction.
HABITS AND HABITAT Diverse habitats so long as there is open ground for males' display, shelter for roosting and edible field-layer vegetation. Polygamous males gather together in spring especially at traditional grounds ('arenas') to display ('lek'), when they strut, crow and fight. Females on the fringes mate with dominant male. Males take no further part in family life.

Red-legged Partridge
■ *Alectoris rufa* L 32–34cm WS 47–50cm

DESCRIPTION Sexes similar. Brown upperparts; warm buff belly; lavender-grey crown, breast and flanks, the last vertically barred white, black and chestnut; white supercilium; black eye-stripe forms a border around white throat, the black breaking down into a bib of streaks. Red bill and legs. Red outer tail feathers show in flight. DISTRIBUTION Natural range is France, N Italy and Iberia. Introduced very widely in Britain from the 18th century onwards. Resident. HABITS AND HABITAT Mainly lowland down to the coast, in open country, arable farmland, pasture and orchards with low vegetation in which it can hide. Eats a wide variety of vegetation. Runs when disturbed rather than flying. Both parents tend young, calling quietly and constantly to keep family together. Commonly reared as a gamebird. Call: loud, harsh *chucka chucka* with variations.

Grey Partridge ■ *Perdix perdix*
L 29–31cm WS 45–48cm

DESCRIPTION A distinctive 'round' gamebird. Orange-brown head, grey neck and breast, and chestnut-barred flanks; broad, inverted horseshoe on white belly; brown upperparts and wings marked chestnut. Rufous outer tail. Blue-grey bill and legs. DISTRIBUTION Widespread from British Isles across NW Europe to S Sweden, S Finland and Baltic states into Russia. Mainly resident. HABITS AND HABITAT Principally agricultural land with rough cover, and on heaths and sand-dunes nearby. When flushed, flies low over a short distance with a whirr of wings. Nests on ground in good cover, with up to 20 eggs. Both adults tend the chicks, which run easily on 1st day. Pair in family until Aug; small flocks (coveys) from then until *c*. Feb. Call: like a rusty gate.

Quail ■ *Coturnix coturnix* L 16–18cm WS 32–35cm

DESCRIPTION Dumpy, partridge-shaped bird, but only as long as a lark. Male: yellow-brown body barred with brown, black and buff. Rufous chest, whitish belly. Buffish throat with blackish centre stripe, and outlined with 2 curved blackish bars from ear coverts. Female: head less well marked.
DISTRIBUTION Summer visitor to S parts of region. Marked annual fluctuations in many areas; good numbers helped by warm S winds. Winters mainly in Africa S of Sahara.
HABITS AND HABITAT Favours dense herbage of natural grasslands and cereal, rape, clover and hayfields. Few nests ever found; population mostly judged by the number of males heard calling the characteristic piping *kwik kwi-ik*, often recorded as *wet my lips*.

Pheasant ■ *Phasianus colchicus* L 53–89cm WS 70–90cm

DESCRIPTION Male and female both have a long tail. Male: larger than female; dark, metallic green head with red wattles; some have a white neck-ring. Body usually chestnut,

covered with black chevrons; tail (over 35cm long) barred black. Female: buffish brown, covered with dark marks, especially on upperparts.
DISTRIBUTION Common and widespread, except S Norway and Sweden. Native of Turkey and Black Sea eastwards; introduced to our region by Romans.
HABITS AND HABITAT Resident. Commonly reared gamebird. Flourishes in farmland, parkland and plantations. Although often seen in the open, it is wary and runs for cover rather than flies. Omnivorous. Flocks in winter. In breeding season, male may have a small harem. Male's call: harsh, loud crowing, often followed in spring by brief, loud wing-flapping.

TOP LEFT: *male*
LEFT: *female*

Corncrake ■ *Crex crex*
L 27–30cm WS 46–53cm

DESCRIPTION Usually seen when flushed and flies a few metres. Buffish body with dark streaks down back, conspicuous chestnut-coloured wings, and dangling legs. A close view shows reddish-brown bars on flanks and under tail, and blue-grey head and breast.

DISTRIBUTION Summer visitor to region except the far N, in suitable habitats. In the British Isles, now certain only in Scottish islands and W Ireland. Winters in S Africa, especially in the E.

HABITS AND HABITAT Prefers grasslands, natural or cultivated, and other rank vegetation, but not wet ground. Marked decline in N and W since c. 1900 due to changes in farming practices. Some increase in parts of Britain thanks to special conservation efforts. Most often recorded by male's rasping *crex crex crex* call, repeated for long periods day and night.

Water Rail ■ *Rallus aquaticus*
L 23–28cm WS 38–45cm

DESCRIPTION Olive-brown upperparts streaked black; grey face and underparts, with wavy black and white vertical stripes on flanks. Wings dark brown. Long red bill, black along top. Long pinkish legs trail in flight.

DISTRIBUTION Across region in suitable habitats S of c. 60°N, with some also in Iceland and SW Norway, but not N Scotland . Many W birds are resident but others migrate. Icelandic birds dependent on habitat around volcanic springs are resident!

HABITS AND HABITAT Reedbeds, swamps and margins of rivers and ponds with plenty of cover from marsh plants. Probes in shallow water for animal and vegetable food. Jerks tail upwards as it walks. Shy, most often noted by its loud, discordant call, used throughout year: grunts that become a high-pitched whistle – likened to pig's squeal.

Spotted Crake
■ *Porzana porzana*
L 22–24cm WS 37–42cm

DESCRIPTION Like a miniature olive-brown Moorhen (below), with white speckling on upperparts and less noticeable white spots on grey-tinged underparts. Conspicuous buff under-tail coverts. Flanks barred black and white. Bill yellow with red spot at base. DISTRIBUTION Summer visitor. Uncommon or rare in W; more widespread around the Baltic but easily overlooked. A few winter in Europe, but most in S and SE Africa.
HABITS AND HABITAT Swamps, bogs and water margins with a thick cover of sedges or similar plants. Secretive and skulking, so distribution and populations hard to measure. May be best noted by its song, a loud, high-pitched *whit*, repeated every second for several minutes, mostly at dawn or dusk; or its call, like a ticking clock.

Moorhen
■ *Gallinula chloropus*
L 32–35cm WS 50–55cm

DESCRIPTION Adult: looks dull black at a distance, with a white line along its side; white under tail is very noticeable as it flicks its tail up. A better view shows brownish-black back and wings, greyish-black underparts. Conspicuous red bill and basal shield, with yellow tip. Yellowy-green legs and feet. Juvenile: dull, dark brown, paler below. DISTRIBUTION Widespread to about 60°N. Resident in the W, but E birds migrate to within rest of range.
HABITS AND HABITAT Territorial on a wide range of still or slow-flowing fresh waters, even on farms and city parks. Feeds mostly on vegetable matter, on or in the water, and often in the open on grassland. Has 2–3 broods a year; earlier young help to feed new chicks. Common call: *curruc*.

■ COOT AND CRANE ■

Coot ■ *Fulica atra* L 36–38cm WS 70–80cm

DESCRIPTION Adult: appears completely slatey black save for white bill and frontal shield. In flight, shows a pale border to edge of wing. Greeny-grey legs and lobed toes. Juvenile: back and wings duller than adult's, and face and underparts whitish.
DISTRIBUTION Widespread; range very similar to Moorhen's (p. 54). Winter visitors to Britain form flocks of >1,000 on several lakes or reservoirs.
HABITS AND HABITAT For breeding, prefers larger waters than Moorhen – lakes and reservoirs with cover around margins. In winter, found on large waters (even floods) with no cover. Very aggressive in territorial disputes, stabbing with bill, and kicking and beating with wings, all on the water. Short, sharp, repeated calls of varying intensity and meaning, e.g. *kowk kowk*.

Crane ■ *Grus grus*
L 110–120cm WS 220–245cm

DESCRIPTION 20% larger than Grey Heron (p. 22). Tall, long-necked and long-legged. At a distance, body and wings appear lead-grey, with long, drooping 'tail' (actually elongated secondary feathers). Closer view reveals a red patch on black crown, and a broad white stripe from eye down side of black neck. In flight, black wing-tips show, and neck and legs are stretched out.
DISTRIBUTION Summer visitor to Germany and Scandinavia eastwards. W populations migrate SW to winter mostly in S Spain; others migrate SE as far as Ethiopia.
HABITS AND HABITAT Typically breeds in swampy woodland, sphagnum bogs and reedbeds. Winters in open country, wetlands, grassland and cultivation. Shy and wary. Gregarious, often in large flocks on migration and at spectacular dancing displays. Bred in Britain until *c.* 1600; since 1981, a small number have bred in East Anglia. Signals its presence with loud, trumpeting calls.

Oystercatcher

■ *Haematopus ostralegus*
L 40–45cm WS 80–86cm

DESCRIPTION Large pied wader.
Summer adult: when on shore, appears
black above and white below; in flight,
shows long white wing bar and white
rump and tail, the latter with a black
terminal bar. Pink legs, long red bill
(8–9cm). Winter adult and juvenile:
white neck bar.
DISTRIBUTION Iceland, British Isles
and France eastwards to Baltic and
Scandinavia. N and E birds in particular
migrate to milder W.
HABITS AND HABITAT Chiefly found on rocky, sandy or estuarine seashores. Nests
and winters in such places. Some also nest inland on shingle by gently flowing rivers.
Gregarious yet wary. Feeds mainly on bivalve shellfish, found by probing and opened by
hammering or stabbing with chisel-like bill. Noisy. Basic call: loud, shrill *kwik kwik kwik*,
often much repeated.

Stone Curlew

■ *Burhinus oedicnemus*
L 40–44cm WS 77–85cm

DESCRIPTION Cryptic sandy-brown
body, heavily streaked darker above,
less so below. Distinctive head pattern:
brown crown, and staring yellow eye
emphasised by white bar above and
dark below. Conspicuous wing pattern:
black primaries with 2 large white spots,
and black–white–black bars across wing
coverts. Short yellow bill with black tip;
yellow legs.
DISTRIBUTION In decline over much
of Europe. Now isolated populations in
SE England and France, eastwards to
Poland. Needs careful conservation of habitat for its survival.
HABITS AND HABITAT In all seasons, occurs on open stony, fallow wasteland with little
ground vegetation. Usually not seen until it moves, with stealthy steps, head and neck
extended. Feeds on invertebrates, mostly from dusk to dawn. Named for its shrill *coor-lee*
call, but not in same family as Curlew (p. 69).

Avocet ▪ *Recurvirostra avosetta*
L 42–45cm WS 77–80cm

DESCRIPTION Unique combination of pied plumage, long blue-grey legs and slender, upturned bill. Predominantly white with a black-capped head, and black wing-tips, coverts and sides of mantle. DISTRIBUTION In suitable habitats from SE England and N France eastwards to S Baltic. Winters locally in Britain and the Netherlands, but more often from Iberia to coasts of Africa, S to Cape of Good Hope. Declined in the 19th century; re-established in SE England from 1947 onwards. Its presence depends on careful control of habitat.

HABITS AND HABITAT Extensive lowland coastal mudflats or brackish lagoons for feeding; nearby dry sandy or muddy flats or low islands for nesting. Has a distinctive feeding method, sweeping its bill methodically from side to side with the open, curved tip just under shallow water, catching prey by touch.

Dotterel ▪ *Charadrius morinellus*
L 20–22cm WS 57–64cm

DESCRIPTION Breeding male: unmistakable. Black crown, white supercilium, dark eye-stripe, whitish face; grey breast, separated from rufous belly by black and white band; belly shades to black, then suddenly white under tail. Wings brown, no wing bar. Breeding female: similar but brighter. Winter adult: plumage pattern muted. DISTRIBUTION Summer visitor to Highlands of Scotland and in suitable habitats in Fennoscandia. Winters in N Africa. HABITS AND HABITAT Breeds on open, flat stony ground, mainly in uplands, with little or no vegetation. Eats mostly insects and spiders. Outside the breeding season forms small flocks ('trips'). Female takes lead in courtship, lays 2–3 eggs, leaves much of incubation and care of chicks to male, and then forms 'hen parties'. Often very tame.

Little Ringed Plover
■ *Charadrius dubius*
L 14–15cm WS 42–48cm

DESCRIPTION Similar to Ringed Plover (below), but slighter, lacks wing bar, yellow eye-ring is noticeable at close range, legs are usually flesh-coloured, and has a different call.

DISTRIBUTION Summer visitor to suitable territory in Britain (mostly England), and from France to states around the Baltic. Winters S of Sahara from W Africa eastwards across the continent.

HABITS AND HABITAT Mainly shallow fresh waters. Nests on banks of shingle, and sparsely vegetated, level sand or gravel borders of rivers. Some recent range extension into gravel pits, sewage works and industrial waste ground; quickly deserts if site changes, so populations fluctuate. Not very gregarious. Feeds on insects from ground or shallow water. 4 eggs laid in a scrape in ground. Call: distinctive *PEE-u*, often repeated.

Ringed Plover
■ *Charadrius hiaticula*
L 18–20cm WS 48–57cm

DESCRIPTION Adult: crown and back brown. Underparts white with broad black breast-ring. Head patterned: broad black from base of bill to lores, black band from eye over rear of forehead, rest is white; chin and around nape white. Bill orange with black tip. Legs orange-yellow. Wing coverts brown, flight feathers black with white bar right across (cf. Little Ringed, above). Tail with brown centre and white edges. Juvenile: lacks black on head; breast-band broken.

DISTRIBUTION Mostly coastal in Iceland and British Isles to Scandinavia and the Baltic. British birds largely resident; E European and Scandinavian birds winter widely from British Isles to W Africa.

HABITS AND HABITAT Breeds on sandy or pebbly shores, and dry sandy or gravelly land by lakes or rivers inland. Winters mainly on sandy or muddy shores and estuaries. Energetic; runs a few metres, then stops and bends to feed on invertebrates; runs again, and so on. Call: melodious *too-LEE* .

Grey Plover ■ *Pluvialis squatarola*
L 27–30cm WS 71–83cm

winter

DESCRIPTION All year shows diagnostic black axillaries in flight, and a short black bill and black legs. Summer adult: upperparts appear silvery grey at a distance but are actually spangled, spotted white and black. Underparts black, divided from crown by broad white band over eye and down side; rump, under-tail and tail white, the last barred black. Winter adult: black face and underparts become speckled pale grey, with indistinct white supercilium; upperparts a dull version of summer.

DISTRIBUTION Breeds outside region in Arctic Siberia. Passage migrant in autumn and spring en route to Africa, many staying to winter in British Isles, France and the Netherlands.

HABITS AND HABITAT Forms small flocks on the coast, often with other waders, to feed on mudflats and sandy beaches on worms, small molluscs and crustaceans, by 'stop-peck-run-stop-peck-run' method common to plovers. Call: far-carrying *tleee-oo-ee*, unlike other plovers.

Golden Plover ■ *Pluvialis apricaria*
L 26–29cm WS 67–76cm

DESCRIPTION At all ages and in all seasons upperparts are distinctively spangled black and gold, and tail is barred. Summer adult: underparts black, divided from crown by white band down side. N birds are more intensely black, while S birds have greyish face. Winter adult: underparts mottled grey-brown, fading to white on belly; off-white supercilium; white underwing (cf. Grey Plover, above).

DISTRIBUTION Breeds in Iceland, N British Isles and across Fennoscandia in suitable habitat. Partial migrant in British Isles; wholly so elsewhere, to W Europe (where populations mix) and S to Iberia and N Africa.

HABITS AND HABITAT Breeds chiefly on upland moors and peatlands, in heather or moor grass. In winter, often forms large flocks to feed on grasslands off the moors, on stubble, arable fields and, especially in hard weather, mudflats. Call: musical *tlui* whistle.

Lapwing ■ *Vanellus vanellus*
L 28–31cm WS 82–87cm

DESCRIPTION Adult: unique, with black and white patterned head and diagnostic long, thin crest; black breast, white belly and pale orange under tail. Black back, metallic green on mantle and wing coverts. Wings broad and rounded, black with white tips. Tail black and white. Juvenile: shorter crest and clear, pale edges to back and coverts.
DISTRIBUTION Widespread across region, but not Iceland. Mainly a summer visitor, wintering in British Isles and W maritime Europe. Very susceptible to cold weather, so moves on freely.
HABITS AND HABITAT Mainly a farmland bird, especially on open arable ground; also favours rushy fields and moorland for breeding. Hard weather will drive flocks to muddy estuaries. Male has a spectacular twisting and turning aerial display in spring. Winter flocks have a characteristic 'leisurely' flight with slow wingbeats. Call: variations on *PEE- wit*.

juvenile, winter

Knot ■ *Calidris canutus*
L 23–25cm WS 57–61cm

DESCRIPTION Rather dumpy and short legged. Breeding adult: upperparts boldly mottled black, chestnut and buff. Face and underparts chestnut-red. Dark eye, black bill and legs. Winter adult: all red lost, becoming dusky grey with intricate dark markings, and white belly, supercilium and rump (speckled dark).
DISTRIBUTION Passage migrant and winter visitor from Canadian high-Arctic islands and Greenland. First arrivals in Jul–Aug still in breeding plumage.
HABITS AND HABITAT In the region, found especially on extensive coastal and estuarine sand- and mudflats, feeding on intertidal invertebrates. Uncommon inland. Very gregarious at this time, often forming dense flocks of several thousand; a flock in flight is one of birdwatching's greatest sights. Call: low-pitched *knut*.

Sanderling ■ *Calidris alba*
L 20–21cm WS 40–45cm

winter

DESCRIPTION Winter adult: distinctive. Most of head and underparts pure white, upperparts pale grey with faint darker markings, darker wings with prominent white wing bar. Dark-centred tail has white sides. Breeding adult: head, upperparts and breast chestnut with dark streaks, especially on back. DISTRIBUTION Breeds in Spitzbergen and Siberia, plus high-Arctic Canadian islands and Greenland. Both populations (proportion not known) are passage migrants and winter visitors to Denmark westwards, and onwards to Africa. Early arrivals are in summer plumage.
HABITS AND HABITAT Well-named characteristic wader of sandy shores. Rushes along beach, legs a blur. Feeds at water's edge, snatching sandhoppers. Gregarious in small flocks. Outside the breeding season, commonly makes a soft flight call: *twick twick.*

Purple Sandpiper
■ *Calidris maritima*
L 20–22cm WS 42–46cm

winter

DESCRIPTION Dumpy with short legs. Summer adult: head and upperparts blackish brown with buff-tipped feathers and purple gloss over back. Breast speckled dark grey-brown, becoming whiter towards belly. In flight, looks dark, relieved only by whitish sides to rump. Dark-tipped bill yellow at base; legs yellow. Winter adult: duller but always dark. DISTRIBUTION In the region, breeds mainly on tundra of Arctic islands, Iceland and Norway; winters in latter 2, plus British Isles (mostly Scotland), Low Countries and France.
HABITS AND HABITAT In winter, almost always on rocky shores, and sometimes on piers and breakwaters. Small parties forage for invertebrates under seaweed and in rock crannies, agilely dodging waves. Dark winter plumage can make them hard to see on wet rocks.

TOP: *summer;* ABOVE: *winter*

Dunlin ■ *Calidris alpina*
L 16–20cm WS 38–43cm

DESCRIPTION In all plumages, has a long, thin white wing bar and noticeable white sides to dark rump. Bill variable in length, more distinctly curved in some. Summer adult: chestnut and black upperparts, grey breast with darker streaks, and broad black patch on otherwise white underparts. Winter adult: loses black patch; head, upperparts and breast ash-brown, streaked darker, rest of underparts white.
DISTRIBUTION Breeds in Iceland, British Isles and Scandinavia, some S of Baltic. Many birds migrate through British Isles to NW Africa. Wintering birds in British Isles and NW Europe are from high-Arctic breeders.
HABITS AND HABITAT Nests on high grassy moorlands with boggy areas (Dartmoor, Devon, is the most S site in the world), and on wet tussocky tundra. Winters on muddy and sandy shores and estuaries, often in large flocks. Commonest call: shrill, rather slurred *treeep.*

Curlew Sandpiper
■ *Calidris ferruginea*
L 18–19cm WS 42–46cm

DESCRIPTION Wader with a striking long, decurved bill. Summer adult: unmistakable, with rich chestnut face and underparts, and mottled black and chestnut upperparts. In flight, shows a white wing bar on dark wings, and white rump contrasting with dark tail. Winter adult: grey above and white below.
DISTRIBUTION A passage migrant in autumn, more rarely in spring, from its breeding grounds in Arctic Siberia.
HABITS AND HABITAT Commonly associates with Dunlins (above) on sand- and mudflats, estuaries and saltmarshes. Numbers fluctuate from year to year; higher numbers may reflect that predators are feeding more on lemmings instead of eggs. Feeds by probing in shallows for worms and other invertebrates. Mostly in small groups. Usual call: soft *chirrip.*

Temminck's Stint

■ *Calidris temminckii*
L 13–15cm WS 34–37cm

DESCRIPTION Breeding adult: mouse-grey above, with dark centres and rufous edges to feathers. Grey breast gorget, streaked brown; rest of underparts white. In flight, shows distinctive white or nearly white borders to rump and tail. Narrow white wing bar. Legs pale, varying in colour from yellowy to brownish (cf. Little Stint, below).Winter adult: as summer, but grey head, breast and upperparts, lightly marked darker. Juvenile: complete brownish-grey breast-band and plain upperparts, without Little Stint's white 'V'.

DISTRIBUTION Breeds in N Fennoscandia and across Siberia. Migrates on a broad front overland to winter in Africa S of Sahara. Only small numbers seen on passage in our region; especially uncommon in the W.

HABITS AND HABITAT Favours freshwater sites, often with low cover by a pool, or saltmarshes. Usually occurs singly or in small parties. Looks longer than Little Stint. Has a habit of 'towering' when flushed. Usual call: short trill, quite unlike Little Stint's.

Little Stint ■ *Calidris minuta*
L 12–14cm WS 34–37cm

DESCRIPTION Tiny wader compared with most. All ages and sexes have black legs. Breeding adult: upperparts markedly chestnut with dark centres to feathers. Sides of breast speckled with orange and brown, rest of underparts white. Dark brown wings with white wing bar. Winter adult: mouse-grey with darker streaks; breast marks greyish. Juvenile: resembles summer adult with white 'V' on mantle (cf. Temminck's Stint, above).

DISTRIBUTION Breeds in far N of Fennoscandia near 70°N and across Arctic Siberia. Winters in Africa S of Sahara. Passage migrant in our region, especially in autumn.

HABITS AND HABITAT Breeds on high-Arctic coastal tundra. On passage, favours mudflats and sandy shores. Gregarious, mostly in small flocks. Dashes about, picking insects and larvae from surface of water or ground. Commonest call: short, low *chit chit*, repeated 2–3 times.

winter

Turnstone ■ *Arenaria interpres*
L 22–24cm WS 50–57cm

DESCRIPTION Breeding adult: black and white patterned head, black breast, and 'tortoiseshell' black and chestnut upperparts. Short, stout black bill. Short orange-yellow legs. Unique flight pattern, with white shoulders, wing bars, lower back and base of tail; the rest looks black. Winter adult: much plainer, with blackish-brown head, breast and upperparts, and white chin and belly.
DISTRIBUTION Breeds in a narrow coastal band around Fennoscandia and the Baltic, and all around the Arctic. Canadian and Greenland birds winter in W Europe, while our region's breeders migrate to NW and W Africa.
HABITS AND HABITAT Usually in small parties on rocky, seaweedy shores. Well named – actively turns over stones and roots in tangled seaweed for a wide range of food. On sandy shores above the high-tide line, repeatedly digs for sandhoppers. Rather quarrelsome in feeding flocks. Flight call: distinctive twittering *kitititit*.

Common Sandpiper
■ *Actitis hypoleucos*
L 19–21cm WS 38–41cm

DESCRIPTION Adult: plain brownish-grey upperparts, streaked breast sharply divided from white underparts; distinctive white gap between carpal bend and side of breast. Brownish-grey coverts and dark brown flight feathers divided by broad white wing bar. Tail projects well beyond wing-tips of perched bird. Legs greenish or brownish. Immature: buff bars on coverts and tertials.
DISTRIBUTION N and W British Isles, Baltic and Fennoscandia; scattered elsewhere. A few winter in mild W Europe, most in tropical Africa.
HABITS AND HABITAT Walks with a constant wagtail-like bobbing. Unique wing action: flicked wingbeats, then a glide, on wings held below shoulder level. Breeds by clear hill streams or lakes. On passage, occurs by lowland waters and saltmarshes, but not open shores. Call: shrill *seep-seep-seep*.

Wood Sandpiper
■ *Tringa glareola*
L 19–21cm WS 56–57cm

DESCRIPTION Like Green Sandpiper (below), but looks slimmer and always lacks Green's black and white appearance in flight. Summer adult: greyish-brown upperwing with much whitish mottling in summer; faintly marked breast; white supercilium; light grey, not black, underwing. Bill dark, legs yellowish; both quite long. Winter adult: upperwing mottling finer. Juvenile: strongly spotted upperparts.
DISTRIBUTION Summer visitor to N Scotland (a few) and Denmark (a few more), but mostly to Fennoscandia and Russia. Migrates on a broad front across Europe, scarcest in W, to tropical Africa.
HABITS AND HABITAT Nest usually on ground in dense vegetation in swamps, bogs and marshes, wet areas in woodland, and wet moorland. On passage, seen mostly on muddy edges of marshes, pools, lakes and sewage works. When flushed, calls an excited, characteristic *chiff-chiff-chiff*.

Green Sandpiper
■ *Tringa ochropus*
L 21–24cm WS 57–61cm

DESCRIPTION In flight, wings appear black above and below (cf. Wood Sandpiper, above) with contrasting white rump and mostly white tail, like a large House Martin (p. 102). Otherwise, upperparts dark olive-brown, lightly spotted; head to breast streaked and spotted dark, sharply divided from rest of white underparts. Greyish-green legs, dark bill.
DISTRIBUTION Summer visitor to E Europe and Fennoscandia. A few winter in mild NW Europe, but commoner on passage to Mediterranean basin and tropical Africa.
HABITS AND HABITAT Breeds in swamps and marshes by old stands of trees. Strangely for a wader, uses the old nest of a thrush, pigeon or squirrel. Non-breeding birds, mostly singly, winter on borders of lakes, streams and drains on marshland and saltmarshes. Call: clear, musical *tooeet-weet-weet* after being flushed.

Redshank ■ *Tringa totanus*
L 27–29cm WS 59–66cm

DESCRIPTION Breeding adult: bright orange-red legs; straight, black-tipped orange-red bill. Olive-brown upperparts, strongly marked with black spots; white underparts with breast and flanks spotted and streaked dark. In flight, shows a conspicuous white rump, tail coverts and wing bars. Winter adult: similar but much plainer. DISTRIBUTION Iceland, British Isles, Low Countries and scattered populations around the Baltic, but mostly in W Scandinavia. Most birds from first 2 areas winter in the British Isles, others stop there on passage to Mediterranean and W Africa.
HABITS AND HABITAT Breeds especially on grassy marshes and wet grasslands in open lowlands and valleys. Male guards territory from a viewpoint on a post. Gregarious on passage and in winter; mainly coastal, especially muddy estuaries and creeks. Normal call: ringing *TEU-hu-hu*, the last 2 notes lower in pitch.

TOP: *summer*; ABOVE: *winter*

Spotted Redshank
■ *Tringa erythropus* L 29–31cm WS 61–67cm

DESCRIPTION Longer legs and bill than Redshank (above). Breeding adult: unmistakable, all dusky black except for narrow white feather edges on upperparts, white rump and narrowly barred black and white tail. Non-breeding adult: no black; ash-grey upperparts, speckled coverts; white below, lightly marked pale grey on breast; noticeable white supercilium.
DISTRIBUTION Summer visitor to Arctic at 65–70°N, but not Iceland. Uncommon passage migrant in NW Europe, where a few winter, most heading to Mediterranean and W Africa.
HABITS AND HABITAT Often feeds by wading and swimming in deeper water than its relatives. Nests in wooded tundra near water. On passage and in winter, seen chiefly singly or in small group on sheltered shores, saltmarshes, lagoons and sewage farms. Characteristic call: clear *chu-it, chu-it*.

Greenshank ■ *Tringa nebularia*
L 30–33cm WS 68–70cm

DESCRIPTION Tall with a long, very slightly
upturned bill. Non-breeding adult: grey above,
white below, with greyish streaks on head, neck
and breast. Wings look all dull black in flight,
in contrast to white lower back, rump and barred
tail. Bill dark grey, legs olive-green. Summer
adult: similar but black spots on back and flanks.
DISTRIBUTION Summer visitor to N Scotland,
much of Fennoscandia and N Russia. Scottish
birds winter in Ireland and W Britain, the rest
in tropical Africa.

juvenile

HABITS AND HABITAT Scottish birds breed in the Flow Country, with
its treeless moorlands, lochs and bogs. Elsewhere, prefers broad mountain
valleys with bogs, lakes and pools, and scattered trees. Nest is on ground close to a marker
– a stone, log or tree stump. Call: ringing *tew, tew, tew*.

Ruff ■ *Philomachus pugnax*
L 20–30cm WS 48–58cm

DESCRIPTION Male ('ruff') noticeably larger
than female ('reeve'). Both sexes have a slightly
decurved bill and yellowish legs, the male's
brighter (even orange), and all ages and sexes
have a large, oval white patch on each side of
dark tail. Breeding male: spectacularly coloured
ear tufts and ruff. Back usually dark and strongly
marked. Underparts mostly black. Female: lacks
ear tufts and ruff; greyish head and underparts,
mottled darker; upperparts have chestnut-edged
blackish feathers. Non-breeding male: similar
to female but clearly larger. Juvenile: buff neck
and breast.
DISTRIBUTION Has suffered through loss of
habitat. Scarce summer visitor across region.
A few winter in W Europe but most in Africa.
HABITS AND HABITAT Breeds on wetlands
with shallow pools and low vegetation. 5–20
males have a traditional dry area for their display
('lek') arena. Dominant males hold territory
in the arena and successfully mate with several
females.

TOP: *male*; ABOVE: *female*

Black-tailed Godwit

■ *Limosa limosa*
L 40–44cm WS 70–82cm

DESCRIPTION Tall, long-legged wader with a long, straight bill (7.5–12cm). Summer adult: head, neck and breast chestnut-red; rest of underparts white. Upperparts grey, mottled with black and chestnut. Unique flight pattern: brown coverts, broad white bar across black secondaries and primaries, white lower rump and base of tail, and black terminal tail band. Legs project well beyond tail. Winter adult: coloured plumage becomes plain, pale grey.

DISTRIBUTION Breeds commonly in Iceland, a few in Britain and many in Germany eastwards below 60°N. Icelandic birds winter in British Isles; others winter in Africa, mainly N of Equator.

HABITS AND HABITAT Breeds in dispersed colonies on marshy moorland, reclaimed grassland, rough pasture and water meadows. In winter, favours mudflats, estuaries and freshwater marshes. Sometimes wades deeply to feed. Very gregarious in non-breeding flocks. Flight call: *wicka-wicka-wicka*.

Bar-tailed Godwit

■ *Limosa lapponica*
L 37–39cm WS 70–80cm

DESCRIPTION Compared with Black-tailed (above) is noticeably shorter-legged, bill is distinctly upcurved, shows no obvious wing bar, and has a black- and white-barred tail. Summer adult: head, neck and underparts chestnut-red; upperparts grey, mottled black and rufous. Non-breeding adult: grey head with whitish supercilium,

winter grey breast; upperparts grey, streaked white; underparts white.

DISTRIBUTION Summer visitor to 65–70°N in Fennoscandia. Passage migrant and winter visitor to British Isles, the Netherlands and Germany, and to Atlantic coasts of Africa.

HABITS AND HABITAT Breeds on peat and heath tundra. After breeding, almost always coastal, on sand or mud. Territorial breeder; later on, very gregarious in large flocks. Often wades up to belly while feeding. Call: *kirruc, kirruc*.

Curlew ■ *Numenius arquata*
L 50–60cm WS 80–100cm

DESCRIPTION Largest wader in the region. Long legs, uniform brown plumage with darker streaks and no distinctive head pattern (cf. Whimbrel, below). Dark brown wing-tips, brown coverts; secondaries and inner primaries paler because of narrow dark and light barring. White rump fades into barred tail. Remarkably long (10–15cm), decurved bill. DISTRIBUTION Breeds widely across the region except Iceland and the central spine of Scandinavia. Winters in the W, southwards to Africa.
HABITS AND HABITAT Breeds on moist moorland, rough grasslands and damp pastures. Outside breeding season, favours mostly mudflats, saltings and neighbouring grassland at high tide. Picks and probes deeply for invertebrates. The most wary of waders. Gregarious, often in large flocks. Characteristic call, heard throughout the year: *CUR-lee, CUR-lee*, 2nd syllable higher than the 1st, and often running into a rich, bubbling sound.

Whimbrel ■ *Numenius phaeopus*
L 40–42cm WS 76–89cm

DESCRIPTION Note head pattern: dark brown crown with buff centre stripe, buff supercilium, dark eye-stripe, and pale cheeks and throat. Most of plumage mottled dark brown, belly and under-tail whitish. No wing bar, but lower back and rump white. Long (6–9cm), curved bill, like that of a small Curlew (above). DISTRIBUTION Summer visitor to Iceland, far N Scotland and islands, Fennoscandia and Russia. Passage migrant to W Europe; winters in Africa S to Cape of Good Hope.
HABITS AND HABITAT Breeds on moorlands, heaths and margins of tundra pools. Outside breeding season, is coastal on muddy and sandy beaches, and rocky shores. Gregarious on migration, usually in small flocks. Most familiar call on migration: rapid, evenly paced *titititititi*, a sure way to tell it from a Curlew.

Woodcock ■ *Scolopax rusticola*
L 33–35cm WS 56–60cm

DESCRIPTION Round-winged, bulky-looking wader, marbled red-brown above and barred buff below. Crown noticeably barred dark and light horizontally, not longitudinally as in Snipe (p. 71); and face less clearly striped. Wonderfully cryptic colours. DISTRIBUTION Not Iceland or N Fennoscandia, but throughout rest of region. Summer visitor to E, wintering in maritime W. Largely resident in British Isles.

HABITS AND HABITAT Solitary, specialised wader of moist woods with ditches and streams or swamps for feeding. In winter, will inhabit more open, damp scrub. Usually seen only when flushed, zigzagging through trees in flight, then suddenly dropping into cover. Male displays by repeatedly flying a circuit ('roding') in the wood with slow wingbeats, calling quiet, snoring notes followed by a loud 'sneeze'. Last note often heard up to 400m away: *tsiwick*.

Great Snipe ■ *Gallinago media*
L 27–29cm WS 47–50cm

DESCRIPTION Hard to tell from Snipe (p. 71) away from known breeding site: stockier and shorter-billed, with heavily barred underwing and underparts; in flight, has dark wing bar bordered each side with white, and bold white corners to tail.

DISTRIBUTION Suffered a big decline from 19th century onwards. Summer visitor. A few now breed in S Baltic states, more in Norway, Sweden and Russia. Migrates on a broad front S to mostly tropical E Africa. Rare on passage in W of our region.

HABITS AND HABITAT Breeds on mountainsides in the N, or wet lowlands elsewhere, even in well-wooded sites with streams or pools. When flushed, is usually silent, flies straight, then drops into cover (cf. Woodcock, above). Generally solitary except at communal breeding display ('lek'), but then is difficult to see because display takes place at dusk or night.

Snipe ■ *Gallinago gallinago*
L 25–27cm WS 44–47cm

DESCRIPTION Medium-sized wader with a long bill (6–7cm). Striped head: dark brown through eye, then buff and dark below, brighter buff and dark above, and dark crown with buff line through it. Dark, richly patterned upperparts, beautiful gold edges to black and chestnut scapular feathers. Bill dark, legs greenish.
DISTRIBUTION Breeds widely across region. British Isles population disperses locally. Other populations migrate to British Isles and NW Europe.
HABITS AND HABITAT Found on open wet ground, marshes, bogs and peat moors with soft soil to probe for invertebrates. In spring, male dives over territory making a 'drumming' or 'bleating' sound as outer tail feathers vibrate. Persistent spring call: *CHIPper, CHIPper, CHIPper*. Towers when flushed, making a harsh call: *scaaap*.

Jack Snipe
■ *Lymnocryptes minimus*
L 17–19cm WS 38–42cm

DESCRIPTION Note head pattern: 2 pale stripes with a dark stripe between, over the eye; dark, unmarked crown. Almost black, glossy upperparts, with scapulars outlined by 2 pairs of golden-buff stripes, very noticeable in flight. Blotched brown and buff breast and flanks, fading to white. Straight, mostly dark bill (about as long as head). Greenish legs.
DISTRIBUTION Summer visitor to N and E Fennoscandia eastwards. Winters widely but sparsely in British Isles and coastal W Europe.
HABITS AND HABITAT Breeds in swamps, among waterlogged birch and willow. In breeding season, male has a diving display with strange calls. In winter, found in swamps, marshes and flooded fields with good cover. Solitary or in scattered groups. When flushed, flies silently, low over a short distance, then drops into cover (cf. Snipe, above). If seen in the open feeding, has a characteristic bouncy gait.

winter

Grey Phalarope
■ *Phalaropus fulicarius*
L 20–22cm WS 40–44cm

DESCRIPTION Summer adult: unmistakable, with brownish-black crown, white patch surrounding eye, all-chestnut underparts, marbled back, and grey wings with narrow white bar. Bill short and yellow with dark tip. Winter adult: utterly different; pale grey above, white below; dark stripe behind eye, not through it (cf. Red-necked, below). Bill brown with dark tip.

DISTRIBUTION Summer visitor to Iceland, Spitzbergen, and all around high Arctic. On passage, usually seen offshore; strong winds bring some inshore in autumn, especially in the SW. Winters offshore in W Africa.

HABITS AND HABITAT The most oceanic wader; in winter plumage, looks like a miniature gull on the water. Swims an erratic course, jabbing for insects. Breeds on marshy tundra or sandier ground in Iceland, with pools. Call: low *wit*.

TOP: *female*; ABOVE: *juvenile*

Red-necked Phalarope
■ *Phalaropus lobatus* L 18–19cm WS 32–41cm

DESCRIPTION Summer adult: distinctive, with slate-grey head and breast, white throat and broad orange-red collar from behind eye to below throat; rest of underparts white. Upperparts dark brown with buff feather edges. Bill black, straight, fine and tapering. Legs blue-grey; feet lobed. Male duller than female. Winter adult: grey above with white feather edges, white below; white face with dark line through eye (cf. Grey, above).

DISTRIBUTION Summer visitor to Iceland, a few in far N Scotland, then from Fennoscandia across Arctic Russia. Migrates mostly overland to winter in Arabian Sea, so uncommon on passage in region.

HABITS AND HABITAT Breeds singly or colonially in bogs with scattered pools. Often has a polyandrous mating system, with females leaving males to incubate eggs and rear the chicks. Feeds on insects, often while swimming. Call: low *cluck*.

Great Skua ■ *Stercorarius skua*
L 53–58cm WS 132–140cm

DESCRIPTION Adult: size of Herring Gull
(p. 77), but heavier-looking with broader wings
and short, wedge-shaped tail. Dark brown over
all save for prominent white patches at base of
primaries, above and below, which 'flash' as wings
beat. Juvenile: more rufous, especially below.
DISTRIBUTION Regular summer visitor only
in Iceland, Faeroes, Orkney, Shetland, and far
NW Scotland and islands. Winters at sea over N
Atlantic, southwards to Brazil and W Africa.
HABITS AND HABITAT Laboured flight, but then very swift and powerful in pursuit of
food (mainly fish when at sea), obtained from gulls, terns and Gannets by piracy. Important
predator of seabirds in breeding season. Breeds on lava plains in Iceland, and on wet grassy
moorland in Scotland. Usually seen singly on passage.

Pomarine Skua ■ *Stercorarius pomarinus* L 46–51cm WS 125–128cm

DESCRIPTION Has 2 colour phases. Adult light phase (common): blackish-brown
cap, back, under-tail coverts; broken grey breast band on white underparts. Wings dark
grey-brown. Ear coverts pale yellow. Adult dark phase (uncommon): nearly uniform dark
brown. 2 central tail feathers elongated and twisted 90° at end, forming a spoon shape.
Juvenile: mostly dark with barred underparts. All have a striking white panel at base
of primaries, which 'flashes' in flight.
DISTRIBUTION Breeds around Arctic from Russia (about 45°E) eastwards. Oceanic
migrant; seen on passage to Atlantic through Baltic and North Sea, wintering N of Equator.
HABITS AND HABITAT Breeds on low-lying tundra with pools, feeding there mostly
on eggs and lemmings. When at sea, eats mostly fish, scavenging from ships and stealing
from other birds. Most likely to be seen on passage from a headland or boat trip.

Arctic Skua ▪ *Stercorarius parasiticus*
L 41–46cm WS 110–125cm

DESCRIPTION The most commonly seen skua, slimmer than a Pomarine (p. 73). Adults have 2 long, central tail feathers and 2 colour phases. Light phase adult: black cap; white cheeks, yellowish at back; light grey breast-band, rest of underparts white; brown-grey upperparts. Dark phase: all sooty grey-brown, save for darker cap. Intermediates also occur. Juvenile: dark bars below and rufous-barred upperparts. All have white wing-patches.

DISTRIBUTION Summer visitor to Iceland, far N Scotland and islands, coastal Fennoscandia and Russia. Pelagic outside breeding season into N and S Atlantic.

HABITS AND HABITAT Colonial nester on tundra, moorland or islands. Aggressive against intruders. Feeds on small mammals, birds and eggs. At sea, eats mainly fish taken from gulls, terns and auks. Swift and agile in pursuit.

Long-tailed Skua ▪ *Stercorarius longicaudus* L 48–53cm WS 105–117cm

DESCRIPTION The smallest skua. Adult: very long (up to 18cm), slender, flexible, central tail feathers. Black crown sharply contrasted with slate-grey upperparts, white breast and pale grey belly. No white wing-patches. Juvenile: grey-brown above, barred whitish; whitish below, barred brown.

DISTRIBUTION Summer visitor to Arctic Fennoscandia and Russia. Pelagic in non-breeding season; wintering area believed to be mostly S Atlantic.

HABITS AND HABITAT Far more oceanic than other skuas, so least often seen on passage. Breeds in tundra zone, on dry ground from coast to well inland up valleys; feeds then especially on lemmings, so numbers fluctuate with food supply. Less piratical than other skuas. Hovers freely and takes food from surface of sea.

Black-headed Gull

■ *Larus ridibundus* L 34–37cm WS 100–110cm

DESCRIPTION Summer adult: chocolate-brown
hood (not black), pale blue-grey upperparts, white
body and tail. Diagnostic broad white band at front
of primaries, which are narrowly tipped black;
underside of primaries dark grey. Bill and legs red.
Winter adult: no hood, just dark spot behind eye.
Immature: yellowish bill and legs, black trailing
edge to wing behind white panel, black-tipped tail.
Juvenile: bright gingery-brown upperparts until
Jul–Aug.
DISTRIBUTION Breeds in Iceland and British Isles,
eastwards to Russia. E and N populations migrate S
and W. Others disperse as far as Mediterranean.
HABITS AND HABITAT Colonial breeder on sand-
dunes, and beside lagoons, lakes and marshes, even
30km from sea. Large numbers winter on estuaries;
often on playing fields, lakes in parks, reservoirs, and
following the plough. Distinctive laughing call.

TOP: *adult summer;* ABOVE: *adult winter*

Mediterranean Gull

■ *Larus melanocephalus*
L 36–38cm WS 92–100cm

DESCRIPTION Summer adult: black head, red
bill and legs, pale grey upperparts and rounded
white wing-tips; at a distance, looks white
except for head. Winter adult: black reduced
to dark mask and grey nape. Immature: dark
face mask, dark brown wings with pale centre
to secondaries, black tail tip; gains adult
plumage by 3rd year.
DISTRIBUTION Has spread from SE Europe
to reach small numbers in England, the
Netherlands, Germany, Denmark. Winters along
coasts of Black, Mediterranean and North seas.
HABITS AND HABITAT Breeds on marshes
and other wetlands near coast, often with
Black-headed Gulls (above). Rarely seen
out of sight of land.
Call: tern-like
kee-er.

TOP RIGHT: *adult summer*
RIGHT: *adult winter*

winter

Common Gull
■ *Larus canus* L 40–47cm WS 110–130cm

DESCRIPTION Smallish bill and rounded head give it a gentle look. Summer adult: white body, darkish grey wings with black primaries and large white tips ('mirrors'). Bill and legs yellowy green. Winter adult: dull grey, speckled head. Immature: almost black primaries, pale coverts, blackish bar across trailing edge of secondaries; white tail has black sub-terminal band. Gains adult plumage in 3rd year.
DISTRIBUTION Not as common as its name suggests. A few in Iceland, British Isles (mostly Scotland and N Ireland) and Low Countries, eastwards to Russia and beyond. Winters on Baltic, and W seaboard southwards to Brittany.
HABITS AND HABITAT Colonial nester by freshwater lochs, on moorland, inshore inlets and marshes. After breeding, favours estuaries, grasslands and sports fields, feeding on invertebrates, e.g. craneflies, earthworms. Call: high-pitched *keee-ya*.

Lesser Black-backed Gull ■ *Larus fuscus* L 48–56cm WS 117–134cm

DESCRIPTION Adult: white head, body, tail and underwings. Mantle and upperwings dark grey, but wing-tips black with white primary tips. Yellow legs and feet. Juvenile: streaked grey-brown body and wings. At all ages and plumages, be careful to separate it from Great Black-back (p. 78). Adult plumage gained in 4th year.

DISTRIBUTION Nests on coastal and lake islets, especially in Iceland, British Isles, Scandinavia and the Baltic.
HABITS AND HABITAT Colonial ground-nesting breeder; some on rooftops of coastal cities. W and Scandinavian birds migrate to winter as far as N and W Africa (many now winter in Britain). Baltic breeders migrate along valleys to the Middle East and E Africa. Omnivorous: eats fish, birds, invertebrates, carrion and rubbish-tip scraps. Calls: *kaw* and *ga-ga-ga*, similar to the Herring Gull's (p. 77).

Herring Gull

■ *Larus argentatus*
L 55–67cm WS 138–155cm

DESCRIPTION Adult: upperparts pale grey, with white leading and trailing edges to wings; black wing-tips with separated white 'mirrors'. Rest of plumage white. Bill yellow with red gonal spot; legs flesh-pink. Juvenile and 1st-winter: dusky, mottled brown with dark brown wing-tips and tail tip; takes 5 years to gain adult plumage. DISTRIBUTION Widespread, common breeder on all coasts. Far N and E populations migrate S and W; others disperse to local estuaries and ports. HABITS AND HABITAT The typical 'seagull'. Most nest on coastal cliffs and grassy slopes on islands, but recently on roofs of buildings. Outside breeding season, seen on estuaries, refuse dumps, fish quays, and following ships not far off shore, and will steal food from seaside tourists. Typical call: loud *kyou-kyou-kyou*.

Yellow-legged Gull

■ *Larus cachinnans* L 52–58cm WS 120–140cm

DESCRIPTION Adult: very similar to adult Herring Gull (above), but note slightly darker grey upperparts (although not as dark as Lesser Black-back, p. 76), smaller white primary spots, red orbital ring (yellow in Herring Gull) and yellow legs. Immature: very like its close relatives, but with paler head. DISTRIBUTION Originally resident only in Mediterranean, NW Africa and coastal SW Europe. Now spread to coasts of France, and first bred in Britain in 1995. Many sedentary, but some disperse as far as Atlantic islands and W Africa. HABITS AND HABITAT Usually seen throughout the year on the coast. Breeds on cliffs and even rooftops. Omnivorous, eating seeds, fish, shellfish and carrion. Only recently given full species status.

Great Black-backed Gull

■ *Larus marinus*
L 64–78cm WS 150–165cm

DESCRIPTION Adult: very large, with all-black back and wings, the latter with white tips and trailing edges. Rest of plumage white. Heavy yellow bill with red spot; flesh-coloured legs (cf. Lesser Black-back, p. 76). Immature: mottled and streaked brown but with paler head; adult plumage gained after 5 years. DISTRIBUTION Breeds on coasts of Iceland, British Isles, Brittany and Denmark eastwards. Most populations resident, simply dispersing to winter along region's coasts and off shore. Far NE populations migrate W and S along coast.

HABITS AND HABITAT Nests singly or in colonies, preferably on small islands and stacks. In breeding season, is a serious predator of shearwaters and Puffins (p. 84), many of latter caught in mid-air. In winter, found on coasts but also scavenges at ports and rubbish dumps. Call: deep, barking *aouk, aouk*.

Little Gull

■ *Larus minutus*
L 25–27cm WS 75–80cm

DESCRIPTION Summer adult: black head with straight border across upper neck, pale grey upperparts, white underparts. White tips to wings, and diagnostic black underwing with white trailing edge. Dark red bill (often looks black); reddish legs. Winter adult: head white save for black spot behind eye and dark cap. Immature: diagonal black wing bar; black outer primaries and terminal bar to tail; adult plumage gained after 2 years.

DISTRIBUTION Summer visitor to the Netherlands (a few), Denmark and S Baltic states, increasingly in Finland eastwards. Seen on passage on North Sea coasts, especially in autumn, wintering southwards to Mediterranean.

HABITS AND HABITAT Breeds in small colonies on lowland freshwater marshes and grasslands with lush vegetation. Mainly coastal on passage, or on nearby lagoons.

Glaucous Gull

■ *Larus hyperboreus*
L 62–68cm WS 150–165cm

DESCRIPTION Can be confused only with Iceland Gull (below), the only other large gull with white primaries. Adult: larger, more powerful looking than Iceland; yellow bill with red spot; flesh-pink legs. Very pale grey upperparts. White wing-tips obvious at rest and in flight. Immature: very pale brown body with faintly barred upperparts, and whitish wing-tips; bill mostly black, fading to pink base.
DISTRIBUTION In our region breeds only in Iceland, Spitzbergen and Arctic Russia, and thence all around Arctic. Disperses to winter southwards to British Isles.
HABITS AND HABITAT Nests on cliffs if available, or stacks, islets and even crags inland. A predator, scavenger or pirate of animal foods (fish, eggs, fledglings). In non-breeding season, most likely to be seen scavenging in a harbour.

Iceland Gull

■ *Larus glaucoides*
L 52–60cm WS 140–150cm

DESCRIPTION Very like Glaucous Gull (above). Adult: mostly white with very pale grey upperparts. Can be distinguished from Glaucous by rounder head, smaller bill, and wing-tips that project well beyond tail of perched bird. Immature: appears plain buff at a distance, but actually speckled and barred, with whitish wing-tips.
DISTRIBUTION Nearest breeding sites to our region are in Greenland, not Iceland! Winters widely across N seas to Iceland, British Isles and Scandinavia.
HABITS AND HABITAT Breeds on coastal cliffs, often among Kittiwakes (p. 80). Feeds then on fish (alive or carrion), eggs and young birds. Non-breeding birds often scavenge on rubbish tips and in harbours – the most likely place to see one in our region.

Kittiwake ■ *Rissa tridactyla*
L 38–40cm WS 95–120cm

DESCRIPTION Looks less 'fierce' than larger gulls, with a rounded head, dark eye, slim yellow bill and black legs, Adult: upperparts blue-grey with black-tipped wings and no 'mirrors'. Juvenile: white head, body and tail, with black spots above and behind eye, rear neck-band and terminal band to tail; black wing-tips, bill, legs and diagonal wing bar; grey forewing and white behind the bar.

DISTRIBUTION Breeds in scattered cliff colonies on almost all coasts except the Baltic. All are pelagic in non-breeding season, S to about 35°N, even weathering huge storms at sea.

HABITS AND HABITAT Builds a compacted nest of weed on a small ledge. Eats fish and invertebrates found off shore on or near the surface. In summer, may be kilometres offshore, and in a large flock at a good food source. Easily recognised at a colony by its call: *kitti-wa-a-a-k.*

Little Tern
■ *Sterna albifrons*
L 22–24cm WS 48–55cm

DESCRIPTION Two-thirds size of Common Tern (p. 82). Note quite long, pointed bill, yellow with black tip; legs yellow. Breeding adult: white forehead, extending over eye; black cap, nape and line through eye. Upperparts pale blue-grey; rest of plumage white, including short, forked tail. Dark leading edge to primaries. Non-breeding adult: white forehead extends to crown, bill dusky black. Immature: more black on primaries, leading edge to wing coverts, and tail tip.

DISTRIBUTION Big decline in recent years in the W due to human disturbance. More common in Germany eastwards to the Baltic and Russia. Winters S to Mediterranean and Africa.

HABITS AND HABITAT Breeds in small colonies, almost exclusively on sandy and shingly beaches. Flies with quick wingbeats; hovers 5–7m above shallow sea, then dives for fish. Call: *kik-kik* or similar.

Sandwich Tern

■ *Sterna sandvicensis*
L 36–41cm WS 95–105cm

DESCRIPTION Differs from
most other *Sterna* terns in
being larger and whiter
looking, with relatively longer
wings and a long black bill
with a yellow tip. Summer
adult: black cap and nape,
with elongated feathers on
cap; upperparts pale ash-
grey, with no dark wing-tip.

Underparts white. Tail forked and short. Winter adult: white forehead (often gained
before leaving breeding area). Immature: as winter adult but with darker wing-tips.
DISTRIBUTION Summer visitor from British Isles and Brittany to S Baltic. Winters
off Africa, as far S as Cape of Good Hope.
HABITS AND HABITAT Nests colonially in scrapes on sandy, shingly coasts; these are
prone to erosion, so colonies move. Call: loud, grating *kirrik*; adult often seen on passage
accompanied by squeaking juvenile.

Caspian Tern ■ *Sterna caspia*
L 47–54cm WS 130–145cm

DESCRIPTION The world's largest tern, nearly as
large as a Common Gull (p. 76), with a large head
and body, and a deep, dagger-shaped blood-red bill.
Summer adult: black crown and nape. Upperparts
and tail silvery grey, underparts white. Primaries
dusky near tip above, and mainly all blackish
below. Tail slightly forked. Legs black. Winter
adult: cap reduced to speckles and black around
and behind eye. Immature: resembles winter adult.
DISTRIBUTION Summer visitor to only a limited
number of colonies around the Baltic. After
breeding, most migrate southwards across Europe
via large rivers and lakes, so rare in the W. Winters
in tropical W Africa, often far inland on rivers and
rice fields.
HABITS AND HABITAT Nests on flat coasts near
shallow, undisturbed waters. Eats mainly fish,
caught by diving from height of several metres.
Main call: loud, heron-like *kraa*.

Common Tern

■ *Sterna hirundo*
L 31–35cm WS 77–98cm

DESCRIPTION Its long (7–12cm), forked tail gave it the old name of Sea Swallow. Summer adult: black cap; white underparts, rump and tail (greyish tint); pale grey upperparts with dark wedge to wing-tips (above and below); carmine-red bill with black tip, bright red legs. Winter adult: forehead and crown white, bill blackish, wings with dusky leading edge. Juvenile: dark primaries, leading edge and bar across secondaries; black soft parts.

DISTRIBUTION Not Iceland or forested highlands, otherwise scattered throughout region. Summer visitor, wintering especially on coasts of W Africa.

HABITS AND HABITAT Colonial breeder. Nest is a scrape on shingle or sand on a rocky islet, coastal heath, gravel pit or even far inland up rivers. Fishes by plunge-diving. Call: high-pitched *keee-yah* and *kik-kik-kik*.

Arctic Tern

■ *Sterna paradisaea*
L 33–35cm WS 75–85cm

DESCRIPTION Hard to tell from Common (above), hence both are known by shorthand name of 'commic' terns. Summer adult: separated from Common by blood-red bill, shorter legs (noticeable at rest), paler, translucent primaries with dark mark only on trailing edge (most noticeable from below), and greyer-tinted underparts. Winter adult: very like winter Common but with black bill. Immature: resembles Common immature.

DISTRIBUTION Iceland, N British Isles and the Netherlands eastwards across region and around Arctic. Summer visitor; mostly coastal, but also many kilometres up some rivers. No other tern breeds in high Arctic. Winters in S oceans, to the pack ice.

HABITS AND HABITAT Has the longest migration of any bird. Nest sites like Common's. Aggressive against intruders, including humans. Call: *kee-arr*, higher in pitch than Common's.

Roseate Tern ■ *Sterna dougallii*
L 33–38cm WS 72–80cm

DESCRIPTION Summer adult: like a 'commic' tern, but has a black bill (with a little red at base), very long, flexible tail streamers and a rosy tint to underparts, although in flight especially looks white overall. Winter adult: white forehead; retains white appearance and long tail.
DISTRIBUTION Summer visitor in our region only to British Isles and W France, a few hundred pairs in all. Other scattered populations on W Atlantic coasts, Indian Ocean and Australasia. Ours winter mostly in Gulf of Guinea.
HABITS AND HABITAT Breeds exclusively on the coast, especially on islands (where it often tunnels into low vegetation), in the entrance to a burrow or on reserves in special nestboxes. An endangered species in the N Atlantic.

Black Tern ■ *Chlidonias niger*
L 22–24cm WS 64–68cm

DESCRIPTION A marsh tern, as distinct from the sea terns (*Sterna*). Breeding adult: unmistakable. Dark slate-grey, blacker on head and underparts, except for white under tail. Underside of wings light grey. Bill black, legs red-brown. Winter adult: wings and upperparts grey, head and body white except for black ear coverts, crown and nape, and dark patch in front of the shoulder. Immature: similar to winter adult but with duskier primaries, secondaries and leading edges to wings.
DISTRIBUTION Breeds in suitable habitat from France to Russia, mostly at 50–60°N (but not British Isles). Winters on coasts of tropical Africa.
HABITS AND HABITAT Breeds on marshes, fens and reedbeds, the nest a floating mat of vegetation. After breeding, seen on passage on lakes, reservoirs and lagoons. Feeds mainly on insects and larvae, mostly by flying at *c.* 2m and repeatedly dipping to the water's surface.

Puffin ■ *Fratercula arctica*
L 26–29cm WS 47–63cm

DESCRIPTION As with all auks, stands upright
but looks long and low as it swims. Summer adult:
black above, white below, white face; wings dark
below. Huge, laterally flattened bill striped blue,
yellow and red – hence nickname 'sea parrot'.
Legs and feet bright orange. Winter adult: bill
reduced in size and colour, face greyish.
DISTRIBUTION Breeds in Iceland, N and W
of British Isles, N coasts of Norway and Finland,
and Spitzbergen. Wanders widely at sea in winter
as far S as the Canaries and W Mediterranean.
HABITS AND HABITAT Lives in colonies, where
birds can be watched closely. Breeds on islands
and sea cliffs in a burrow 1–2m long, excavated
by the birds and often used again in other years.
Eats fish; chicks are fed sandeels, several at a time
brought in bill. Uses wings to 'fly' underwater.

Black Guillemot ■ *Cepphus grylle* L 30–32cm WS 52–58cm

DESCRIPTION Summer adult: all black (including slim bill), save for large white oval
wing-patch and red legs. Winter adult: white head, neck and underparts, with greyish eye-
patch and crown; upperparts blackish, barred with white; black wings retain white oval.

DISTRIBUTION Breeds on
coasts of Iceland, islands and
mainland of Scotland and
Ireland, and in Scandinavia,
including around the Baltic.
Disperses in winter, moving
less far than any other auk.
HABITS AND HABITAT
Nests in small colonies
in rock crevices, under
boulders and even in holes
in harbour walls. Eats a
large variety of fish and
crustaceans, caught by
diving from the surface,
N birds travelling as far
as the edge of the pack
ice to feed.

Guillemot ■ *Uria aalge*
L 38–41cm WS 64–70cm

DESCRIPTION Summer adult: head, neck and upperparts chocolate-brown (looks black in poor light); underparts and trailing edge to secondaries white. Winter adult: neck and face below and behind eye white. Juvenile: similar to winter adult.
DISTRIBUTION Breeds in N and W British Isles, Iceland and coasts of Fennoscandia. Disperses in various directions in winter, although many stay near their nest sites; 1st-year birds go furthest, but no further than Portugal.
HABITS AND HABITAT Gregarious at all seasons. Prefers nesting on cliffs, laying 1 pear-shaped egg on bare rock, its shape helping it to roll around, not off, ledge. Young are fed by male and leave ledge before they can fly, achieving this after 8–10 weeks. Much threatened by oil pollution at sea.

Razorbill ■ *Alca torda*
L 37–39cm WS 63–68cm

DESCRIPTION Summer adult: thick-set neck, large black bill with white vertical line part way along, and white line from eye to bill. Head, neck and all upperparts black, save for white trailing edge to secondaries. All underparts white. Winter adult: only forehead, crown and nape black; rest of face and underparts white.
DISTRIBUTION Breeds in Brittany (a few), Faeroes, British Isles, Iceland and Fennoscandia, including the Baltic. Winters on inshore waters around breeding areas, N birds and juveniles moving furthest, as far as Morocco.
HABITS AND HABITAT Often found with Guillemots (above) on breeding cliffs, but lays single egg on a wider ledge under an overhang, in a crevice or among boulders at base of cliff. Chick jumps to sea when only two-thirds grown, and is cared for by male.

Little Auk ■ *Alle alle*
L 17–19cm WS 40–48cm

DESCRIPTION Small, dumpy, pied bird with stumpy black bill. Summer adult: underparts, trailing edge of secondaries and fringes of scapulars white, the rest black. Winter adult: gains white on neck and throat. DISTRIBUTION In our region, breeds only on Iceland (a few), and Spitzbergen, Franz Josef Land and Novaya Zemlya (millions!). Disperses after breeding to winter at sea, but usually N of 55°N.

winter HABITS AND HABITAT Nests in huge colonies in rock crevices and scree. Very gregarious. Feeds on small crustaceans in the plankton, resulting in red droppings at roosts on ice floes. Most likely to be seen in our region off North Sea coasts after heavy weather has driven birds inshore.

Rock Dove ■ *Columba livia*
L 31–34cm WS 63–70cm

DESCRIPTION The ancestor of all domestic pigeons. Wild individuals are grey, with a darker head, breast, flight feathers and terminal tail bar. Has 2 black bars across the secondaries, visible at rest and in flight. Green and purple sheen on neck. White rump shows well in flight. Some feral pigeons in our towns and cities are similar, but many have mottled or speckled plumage, with white or brown. DISTRIBUTION Very hard to know the original extent of the species' distribution, because it is now so widespread in the region as a feral bird.

HABITS AND HABITAT Traditional home is believed to be cliffs and rocky landscapes on or near the coast or far inland. Nests in colonies (hence the ease with which it was tamed into dovecotes). In our region, the best places to find birds in wild plumage are far N Scotland and W Ireland. Song: *oor-roo-coo*.

Stock Dove ■ *Columba oenas*
L 32–34cm WS 63–69cm

DESCRIPTION Mostly blue-grey. Broad blackish tips and trailing edges to wings show above and below; 2 narrow black bars on greater coverts; black band to end of tail. Glossy green sides to neck; breast vinous pink. Legs bright pink, bill horn with white cere.

DISTRIBUTION Widespread in British Isles (not N Scotland) and rest of region (not Iceland), mostly below 60°N. Largely resident in the W; migrants from Germany eastwards, wintering in mild W southwards to Iberia.

HABITS AND HABITAT Almost everywhere where mature trees are close to open ground

for feeding on seeds, e.g. copses, ancient parkland, city parks. Nests in a hole in an old tree, rock crevice, ruined building or nestbox. Song: *OOoo-er*, repeated quickly up to 10 times.

Woodpigeon
■ *Columba palumbus*
L 40–42cm WS 75–80cm

DESCRIPTION Another grey pigeon with a green and purple sheen on its neck, and a pink breast. Adult: distinguished by white patch on each side of neck, unique white bar across (not along) wing coverts, and grey tail with black terminal band above and grey, white and black below. Bill yellow, legs pink. Juvenile: lacks neck bands.

DISTRIBUTION Widespread throughout except Iceland,

mostly below 65°N. Largely resident in the W; populations increasingly migratory to the E and N. Can form very large flocks.

HABITS AND HABITAT Wooded country bordered by fields, where it feeds on weed and crop seeds. Increasingly in parks and cities. Needs fresh water for drinking, as do other pigeons. Distinctive, deeply undulating display flight, with wing claps at top of each rise. Song: *coo COOO coo coo-coo-cuk*.

Collared Dove

■ *Streptopelia decaocto*
L 30–33cm WS 47–55cm

DESCRIPTION Dainty dove with a long tail. Adult: mostly pale pinkish buff with dusky-brown primaries. Upper tail brown with white tip, this especially broad at corners; white with black base below, very noticeable in flight. Thin black bar at sides of neck. Juvenile: lacks neck bar.
DISTRIBUTION Throughout region, mostly S of *c.* 60°N, but spreading N. Mostly resident; juveniles disperse.
HABITS AND HABITAT Found near human habitation, especially farms, and gardens with bird tables in towns and villages. Amazing spread from Balkans in 1930s; first bred in SE England in 1955, Ireland in 1959 and Finland by 1966. In display flight, rises steeply then glides down, often calling a harsh *kurr* as it lands. Sings repeatedly from open perch: *coo-OO-cuk*.

Turtle Dove ■ *Streptopelia turtur* L 26–28cm WS 47–53cm

DESCRIPTION Beautifully marked small dove. Adult: head blue-grey with black- and white-striped neck patch, breast pink, back brown and grey, feathers marked with black centres; diamond-shaped tail edged black and white. Wing coverts orange-brown with black centres, flight feathers brown-black with blue-grey between them and the coverts.

Blackish bill, dark pink legs. Juvenile: browner and lacking neck patch.
DISTRIBUTION Summer visitor to S and E Britain, and to France eastwards to Russia, but not Scandinavia. Winters in Africa in savannah S of Sahara.
HABITS AND HABITAT Found in open woods, copses and farmland with thick hedges. Feeds on the ground on wild and cereal seeds. Big decline in recent years. Large flocks seen on migration. Song: purring *turr-turr-turr*.

Cuckoo ■ *Cuculus canorus*
L 32–34cm WS 55–60cm

DESCRIPTION Small head, decurved bill, slim body, long tail. Adult male: head, breast and upperparts slate-grey; wing-tips darker and pointed. Rest of underparts white, barred with brownish-black bars. White-tipped tail droops below wings at rest. Adult female: buff on lower breast; also has a rare rufous phase, barred with black. Juvenile: has plain grey and rufous phases, both with a white patch on nape.
DISTRIBUTION Summer visitor throughout region, even the Arctic, but not open tundra. Winters in Africa, mostly S of Equator.
HABITS AND HABITAT Infamous parasitic nester. Promiscuous, female laying up to 25 eggs, 1 per nest; Reed Warbler (p. 124), Dunnock (p. 110) and Meadow Pipit (p. 105) are the main hosts. Flies with wings held below shoulder level. Often mobbed by small birds. Male's call: famous *cuc-oo*. Female's call: bubbling.

Tawny Owl ■ *Strix aluco*
L 37–39cm WS 94–104cm

DESCRIPTION Large, round head and broad, rounded wings. Black forward-facing eyes in a greyish-brown face mask. Most of plumage rufous brown in W birds, greyer in E birds, streaked above and below; wings have 2 broken white bars.
DISTRIBUTION Widespread resident, but not Iceland, Faeroes, Scottish islands or Ireland, and mostly in S in Scandinavia.
HABITS AND HABITAT Nests in tree-holes, old nests and special nestboxes, in woods, copses, parkland and even town suburbs. Nocturnal. Roosts in thick cover. If found by a small bird, its alarm call soon gathers others to mob the owl. Hunts mainly for rodents at night on silent wings. Indigestible bones and fur are ejected in a pellet. Male calls *hoo-hoo*, both birds *ke-wick*; together this is the traditional *tu-whit-to-woo*.

Eagle Owl ■ *Bubo bubo*
L 60–75cm WS 160–188cm

DESCRIPTION Largest owl, bigger than a Buzzard (p. 44). Upperparts warm brown, thickly marked with black; underparts yellowish brown with black streaks, especially on breast. Brownish-grey face mask with large orange eyes, and 2 often upstanding ear tufts.
DISTRIBUTION Large decrease since the 19th century, mainly due to persecution, but much good protection in recent years. Resident. A few in Britain (probably feral 'escapes'), Low Countries (reintroduced), and Germany to Poland and Fennoscandia, around Baltic; more common from Finland eastwards.
HABITS AND HABITAT In wilderness, scattered populations occur remote from persecution, in rocky mountains and forest with cliffs (favoured nest site). Nocturnal. Eats mammals to size of hares and birds to size of wildfowl. Song: deep, repeated OO-*hu*, OO-*hu*, often audible at >1km.

juvenile

Snowy Owl ■ *Bubo scandiacus*
L 53–66cm WS 142–166cm

DESCRIPTION Male: pure white, usually with a few dark spots on primary tips and coverts. Yellow eyes. Legs and toes fully feathered. Female: up to 20% larger than male; white, with underparts and upperparts barred dark brown.
DISTRIBUTION Breeds beyond the tree-line, N of 65°N in numbers fluctuating according to food supply. Can live in sub-zero temperatures, but lack of prey and winter darkness send birds S, rarely even as far as British Isles.
HABITS AND HABITAT Breeds on open, mostly lowland tundra, with dry hummocks and tussocks for nest site, and similar sites and rocks for lookouts. At tundra sites, eats lemmings and voles almost exclusively; elsewhere, birds and mammals as large as a hare. Call: loud, booming *hooo*.

Great Grey Owl ■ *Strix nebulosa* L 65–70cm WS 134–158cm

DESCRIPTION Large owl with long, rounded wings and relatively long tail. Round face with distinctive white 'half-glasses' framing eyes and resting on yellow bill. Dusky-grey plumage, paler on belly, well streaked with dark brown. Broad black terminal band on tail. In flight, note large buff patch at base of primaries.

DISTRIBUTION Resident, but nomadic when food source fails. Fennoscandia, eastwards from *c.* 20°E and S of 70°N.

HABITS AND HABITAT Mainly dense lowland forest, hunting over nearby moors, bogs and clearings. Crepuscular, but hunts in broad daylight in winter and for young. Watches for prey (mostly voles) from a perch. Nests in old, large bird nests. Song: deep, booming, evenly spaced *hoo-hoo-hoo*, in series of usually 8–12.

Ural Owl ■ *Strix uralensis*
L 60–62cm WS 124–134cm

DESCRIPTION Large, round-headed, dark-eyed, long-tailed owl. Upperparts buffish to grey-brown, much streaked with dark brown; underparts greyish white, evenly streaked with dark brown. Barred wings have no pale panel.

DISTRIBUTION A few in SE Norway, but mostly eastwards from Sweden and Finland, and a few S to Poland. Resident.

HABITS AND HABITAT Nocturnal. Mostly in old boreal forests with open ground for hunting small mammals and birds. Hunts from a perch. Easily identified if seen well, but in fleeting view can be mistaken for Great Grey (above) or Short-eared (p. 93) owls, or even a Goshawk (p. 45). Numbers fluctuate with mammal numbers. Very aggressive in defence of nest in tree-hole or old bird's nest. Alarm call: sharp *kark*. Territorial hoot: a deep cooing sound, *wo-ho* [2–4-second gap] *wo-ho uh woho*.

Tengmalm's Owl ■ *Aegolius funereus*
L 24–26cm WS 54–62cm

DESCRIPTION Appears large-headed. Upperparts and tail brown with rows of white spots; underparts whitish, blotched with light brown. Wings rounded. Striking face pattern: white facial disc outlined in black, raised white eyebrows, black smudges above and below yellow eyes; often recorded as 'Looks astonished!'
DISTRIBUTION Mostly resident. A few in France, Low Countries and Germany, more in Fennoscandia to within Arctic Circle, and around the Baltic eastwards.
HABITS AND HABITAT Breeds in mature forest and clearings. Nocturnal. Nests in a tree-hole. Hunts from perch for mice, voles, shrews and small birds. Best found by locating singing male, whose rapid, deep whistles, *po-po-po-po-po* (in a series of 5–8), are audible in still air up to 1–2km away.

Long-eared Owl ■ *Asio otus*
L 35–37cm WS 90–100cm

DESCRIPTION Stands tall when alarmed, looks short and fluffed up when at ease. Wings long; dark carpal patch, yellowish-buff area at base of primaries, wing-tip evenly barred. Upperparts a rich mixture of browns and greys, heavily streaked and barred with dark brown. Underparts buff, heavily streaked with brown. Facial disc orangey buff, outlined blackish brown; long white eyebrows and base of bill; glowing orange eyes; head capped by long ear tufts.
DISTRIBUTION Widespread, but not Iceland and less so in W Britain and France, N to about 65°N. Mostly resident but N birds migrate S and W in winter.
HABITS AND HABITAT Deciduous and coniferous woods, copses and plantations with nearby open country. Nests in old bird's nest. Preys mainly on small mammals. Song: repeated series of deep, evenly spaced *oo* sounds.

Short-eared Owl ■ *Asio flammeus*
L 37–39cm WS 95–110cm

DESCRIPTION Similar to Long-eared (p. 92), but paler face mask, short ear tufts usually invisible, yellow eyes outlined in black and white, streaked breast but whiter belly, and upperparts golden brown, heavily marked with black. Wing has big yellowy-buff primary patch like Long-eared, but wing-tip looks as if tipped black, and trailing edge of wing is narrowly white.
DISTRIBUTION Breeds in Iceland, N Britain, Low Countries, N Germany, Denmark, S Sweden, Norway, Finland and N Baltic states eastwards. N and E populations migrate S and W to W of region, including S England, W France and Ireland.
HABITS AND HABITAT Nests on ground in optimal habitat of tundra, moorland and young plantations. Hunts at all times of day and night, quartering the ground for voles especially. Song: a series of far-carrying hoots given in flight.

Barn Owl ■ *Tyto alba*
L 33–35cm WS 85–93cm

DESCRIPTION British and French birds have mottled grey and buff upperparts and wing coverts, and pure white underparts; E birds have mainly grey upperparts and yellowish-orange underparts, finely speckled with brown. In flight, underwing is white, and flight feathers are buff with faint darker bars. Heart-shaped facial mask white (less bright in E birds), thinly outlined in dark brown; eyes black.
DISTRIBUTION Widespread across region, but no further N than Scotland, Denmark and Latvia. Sedentary.
HABITS AND HABITAT An owl of farmland, rough grassland, hedges and copses. Crepuscular and nocturnal. Nests in tree-holes, ruined buildings and barns. Serious declines in many countries due to loss of nest sites. Recovers where nestboxes are provided. At dusk, looks ghostly, quartering slowly and low, and hovering over open ground in search of small mammals. Call: drawn-out screech, *shreeee*.

Little Owl ▪ *Athene noctua*
L 21–23cm WS 54–58cm

DESCRIPTION Upperparts and wings dark brown, all spotted with white, or buff on flight feathers; tail barred light and dark. Underparts white, liberally streaked dark brown, especially on breast. White facial disc, yellow eyes and bill, and dusky cheeks give it a fierce look.
DISTRIBUTION Resident in England, Wales and SE Scotland, eastwards to Poland; none in Iceland, Ireland or Fennoscandia.
HABITS AND HABITAT Breeds in farmland, old orchards, river valleys with pollarded trees and old quarries, wherever there is a nest site in a hole. Perches erect on long legs, looking for prey of small mammals and a large variety of insects. When alarmed, bobs rapidly. Commonest call: mewing *kee-ew*.

Nightjar
▪ *Caprimulgus europaeus*
L 26–28cm WS 57–64cm

DESCRIPTION In poor light appears to be dusky brown all over, except that male has white spots near base of outer primaries and outer tips of tail; both sexes have white throat. The plumage is actually cryptic: an intricate pattern of blotches, streaks and bars, of silvery grey, black, buff and rufous. Bill small, gape large.
DISTRIBUTION Summer visitor to suitable habitats in British Isles (not N Scotland) and France eastwards to S Scandinavia, Baltic states and Russia. Winters in Africa S of Sahara away from forest.
HABITS AND HABITAT Found in open woodland, moorland, new plantations, heath and sand-dunes. Crepuscular, aerial insect feeder; silent flight on long wings with many twists and turns, and hovering. Male has a trilling song, a bit like breathing a purring *rrrrrrrrrrrrr*, sometimes for several minutes. Also *coo-ik* from both sexes.

Hoopoe ▪ *Upupa epops*
L 26–28cm WS 42–46cm

DESCRIPTION Unmistakable in flight or on ground: long (5–6cm), tapering, decurved bill; large black- and white-tipped crest, raised when excited; pinkish-brown plumage from crest to flanks, brightest on head and crest; black- and white-barred wings and tail. Tail quite long, wings rounded. DISTRIBUTION Uncommon breeder in our region (commoner further S), from France to Germany and, more commonly, Poland eastwards. Summer visitor. A few overshoot further N in spring. Small numbers winter in S Spain and N Africa, but most S of Sahara. HABITS AND HABITAT Open country with scattered mature trees for nest-holes, such as orchards and pollarded willows by rivers. Feeds on ground, jabbing with its bill for insects and larvae. Song: quiet, low-pitched, far-carrying, trisyllabic *poo-poo-poo*.

Kingfisher ▪ *Alcedo atthis*
L 16–17cm WS 24–26cm

DESCRIPTION A short-tailed, long-billed jewel with bright blue upperparts and orange underparts. Tone of blue changes with the light – back and rump can flash with a pale, intense sheen. Head pattern distinctive: crown, moustache and nape blue; lores and ear coverts orange; throat and patch behind ear coverts white. Bill all black in male, much of lower mandible reddish in female. DISTRIBUTION Through much of region except N Scotland and most lands N of 60°N. In winter, birds in Finland and from Russia to Poland migrate W to ice-free waters. HABITS AND HABITAT Freshwater rivers, streams, canals and lakes, with still or gently flowing water in which to fish. Territorial in breeding season. Digs a tunnel with an egg chamber at end, usually 45–90cm long, in a bank generally overlooking water. Call: shrill *ti-tee*.

Swift ■ *Apus apus*
L 16–17cm WS 42–48cm

DESCRIPTION All blackish brown with a whitish throat. Long, narrow, scythe-like wings. Short, forked tail. Sometimes misidentified as a Swallow (p. 103), but that family has white underparts.
DISTRIBUTION Summer visitor throughout much of region, but few in far N of Scotland, and not on N islands, Iceland or much of far N of Fennoscandia. Winters in southern Africa.
HABITS AND HABITAT Natural nesting habitat of cliffs and crags has been almost wholly replaced by holes and crevices in buildings. Wanders widely for food, anywhere it can hunt insects in the air, so seen over a wide variety of habitats. Fast, agile flight, often seeming to have alternate wingbeats. Very noticeable at nest site, where several pairs may race around screaming *sreeeeee*.

Black Woodpecker
■ *Dryocopus martius* L 45–57cm WS 64–68cm

DESCRIPTION Size and colour of a crow, with a horn-coloured, dagger-shaped bill. Male: glossy black except for red forehead and crown Female: black like male with red spot on nape.
DISTRIBUTION From E France eastwards, to well within Arctic Circle, but not British Isles or N islands.
HABITS AND HABITAT Resident. Needs unbroken mature forest, with large enough trees in which to bore a nest-hole. Feeds on larvae and adult wood-boring beetles, hammered out of tree trunk or stump; also feeds on ground on ants and their larvae. Flies straight with floppy wingbeats, but without other woodpeckers' undulations. Both sexes drum, like a machine-gun burst, to mark territory or attract a partner. Flight call: *krük krük krük*. Song: 10–20 high-pitched, laughing *quee-quee-quee…*.

Green Woodpecker

■ *Picus viridis* L 31–33cm WS 40–42cm

DESCRIPTION Adult: yellowy green above and pale greeny grey below, with lightly barred flanks; in flight, shows bright yellow rump and cream bars across dark brown primaries. Distinctive head pattern: red crown and nape, black around eye, and black moustache, the male's with a red centre. Juvenile: white-spotted upperparts and brownish-black bars on face and underparts.
DISTRIBUTION Not Ireland, N Scotland or Iceland. Otherwise widespread resident eastwards, and N to 60°N and slightly beyond in Norway and Sweden.
HABITS AND HABITAT Open deciduous woodland, orchards and parkland. Very specialised woodpecker, feeding mostly on ground on ants and their pupae, digging into ant-nests and extending its very long, sticky tongue. Drums rarely. Call: laughing *cue-cue-cue*.

Grey-headed Woodpecker ■ *Picus canus* L 25–26cm WS 38–40cm

DESCRIPTION Adult: similar to, but smaller than, Green Woodpecker (above), but note different head pattern and lack of barred flanks on pale grey underparts. Head is almost completely plain grey; both sexes have a narrow black moustache, and male has a red patch on forehead. Juvenile: very similar but duller.
DISTRIBUTION Resident, but not Iceland, British Isles, Denmark, or most N parts of France to Poland. More widespread further S in France eastwards; some in S Scandinavia.
HABITS AND HABITAT More catholic in its habitat than the Green: old timber by rivers and lakes; open, mature deciduous forest; and open coniferous forest in uplands. Forages on trees and on ground for ants and other insects. Call: usually 5–8 whistles, *kee-kee-kee*, dropping in pitch and slowing down.

Great Spotted Woodpecker
■ *Dendrocopus major* L 22–23cm WS 34–39cm

DESCRIPTION The 1st of 3 pied woodpeckers.
Black upperparts with white scapulars forming
big ovals, 4 narrow white wing bars; white
underparts except for crimson-red vent.
Whitish forehead, black crown (rear of male's
crown is red), and black moustache from base of
bill around white cheeks to meet at black nape;
robust black bill.
DISTRIBUTION Resident throughout region
except Ireland, Northern Isles, Iceland and far
N of Fennoscandia.
HABITS AND HABITAT Commonest pied.
Breeds in all kinds of woodland. Eats mainly
insects and larvae, plus many coniferous seeds
in winter; if seed source fails in the N, many
erupt S and W. Uses stiff tail to help control
head-first climb up tree, and has bouncing flight
common to all pieds. Drums with bill in spring
for *c*. 0.5 seconds, dying away at end, on a dead
branch. Call: sharp *kik*.

Middle Spotted Woodpecker
■ *Dendrocopus medius* L 20–22cm WS 33–34cm

DESCRIPTION Similar to Great (above) but
slightly smaller, with white face, buff forehead
and bright red crown. Incomplete moustache
does not reach bill or nape, but does extend
to side of breast as in Great, and merges into
streaked, buff-tinted underparts, which fade to
pink under tail. Bill short.
DISTRIBUTION Scarce resident from France to
Poland, Belarus and Russia.
HABITS AND HABITAT Inhabitant of ancient
deciduous forest, especially with Hornbeam
Carpinus betulus and oaks, so very scattered
populations owing to forest destruction and
unhelpful modern forest management. Rarely
drums, so listen for call throughout year, a rapid
rattle, *kik kekekekek*, 1st note pitched higher
than rest.

Lesser Spotted Woodpecker ■

Dendrocopus minor L 14–15cm WS 25–27cm

DESCRIPTION Only two-thirds the size of Great (p. 98). More quickly identified than other pieds by bold black and white barring on wings and back. Male's crown is red, female's black. Off-white underparts, including under tail, lightly streaked with black. Buff-tinted white face; moustache streak does not meet black nape.
DISTRIBUTION Resident across region except Ireland, N Britain, N islands and oceanic edge of Fennoscandia.
HABITS AND HABITAT Breeds throughout in deciduous woodland. Often feeds high in the canopy, even among small twigs like a tit. Nest built in decayed wood, usually much higher than other woodpeckers. Drums in spring, with a longer series than Great, *c.* 1.5 seconds, and not dying away at end. Most frequently heard call: series of piping notes, *pee-pee-pee-pee-pee*.

Wryneck ■ *Jynx torquilla*

L 16–17cm WS 25–27cm

DESCRIPTION Aberrant woodpecker: long (not stiff) tail, and basically brown and grey plumage, looking like a large warbler. Close view reveals detailed pattern of grey crown and back, with black band down centre and black eye-stripe; rest of head and upper breast buff, fading to white belly, all finely barred black; tail grey with broad,

dark brown bars, wings brown with darker and ochre bars. Short, pointed bill.
DISTRIBUTION Summer visitor, but not British Isles (except on passage), N islands or N Fennoscandia. Scarce from Brittany to Denmark. Winters in Africa S of Sahara.
HABITS AND HABITAT Breeds in open woodland, orchards and copses. Eats mainly ants and their larvae. Picks up food with rapid flicking of long tongue. Nests in an existing hole in tree, bank or nestbox. Song: deliberate series of *kwee-kwee-kwee*, more musical than similar by Lesser Spotted Woodpecker (above).

Skylark ■ *Alauda arvensis*
L 18–19cm WS 30–36cm

DESCRIPTION Terrestrial, streaky grey-brown bird, which walks rather than hops. Streaks on breast sharply divided from rest of white underparts. In flight, shows white trailing edge to broad wings, and white outer tail feathers. White supercilium and blunt-ended crest.

DISTRIBUTION Widespread through most of region except Iceland, high Arctic and spine of Norway. Mostly resident; N and E populations migrate to winter in milder W. Widespread decline in W owing to agricultural intensification.

HABITS AND HABITAT Breeds in open farmland, heaths, moors, coastal dunes and even airfields. Rarely perches on trees. Most often noticed when male flies up steeply, singing very varied, melodious, non-stop outpouring to 'song post' in the sky, *c.* 100m up, for several minutes, followed by slow descent. Call: a liquid *chirrup*.

Crested Lark
■ *Galerida cristata*
L 17cm WS 29–38cm

DESCRIPTION Similar size to Skylark (above) but with shorter tail and more uniform, duller plumage, and looks more dumpy. At all times, even when it is depressed, the pointed crest is very noticeable. Stout, slightly decurved bill. In flight, shows no white trailing edge to wings and buff outer tail feathers.

DISTRIBUTION Resident. Scattered populations across Continental part of region, so not British Isles, Norway, Sweden or N islands. Decreased in much of region.

HABITS AND HABITAT Open country, less so on grassland and more fond of dry ground than Skylark, even roadsides, railways and wasteland. Often allows close approach of observer. Usual call: melodious, sad-sounding *whee-whee-weeoo*, with variations.

Woodlark ▪ *Lullula arborea*
L 15cm WS 27–30cm

DESCRIPTION More richly coloured than Skylark
(p. 100). Erectile crest; short tail with white tips;
no white trailing edge to wings, but does have
whitish wing bar, which shows as a distinctive
white mark next to dark brown on folded wing.
Whitish supercilia meet in a 'V' at back of neck.
DISTRIBUTION S England and across region,
mostly S of 60°N. N and E birds are migratory
to the S and W.
HABITS AND HABITAT Has decreased widely
owing to loss of dry grassland and heath. Also
on open forest (especially pine), scrubby
hillsides and scattered trees. Less gregarious
than Skylark. Dispersing, migrating birds call
in flight, a distinctive *tit-loo-eet*. Male delivers
song fluttering over a wide area, a melodious, sad-
sounding series of notes, falling in pitch, always
with characteristic *EE-lu-EE-lu-EE-lu* phrase.

Shorelark
▪ *Eremophila alpestris*
L 14–17cm WS 30–35cm

DESCRIPTION Most easily
recognised lark in region.
Upperparts mottled pinkish
brown, wings dark brown,
underparts white with russet
streaks on flanks, tail black with
brown centre and white edges.
Distinctive head pattern: pale
yellow face, forehead and throat;
reddish-brown crown with (on
male) narrow, erect black 'horn'
over each eye (hence its North American name of Horned Lark); black lores and stripe
below eyes; black gorget across upper breast.
DISTRIBUTION In our region breeds only in Arctic Fennoscandia and Russia, and winters
on North Sea coasts.
HABITS AND HABITAT Breeds above tree-line on mountains, tundra and barren steppes.
Most likely to be seen in small winter parties on high-tide lines of seaweed on the seashore,
saltings or stubble, where it walks picking up seeds. Call: shrill *tseep* or *tsee-sirrp*.

Sand Martin ■ *Riparia riparia*
L 12cm WS 27–29cm

DESCRIPTION Smaller than its 2 relatives described here. Uniform brown upperparts; white underparts with broad, well-marked brown breast-band. Forked tail. Dusky underwing.

DISTRIBUTION Summer visitor throughout the region except Iceland and high Arctic. Winters in Africa S of Sahara.

HABITS AND HABITAT Very closely attached to water, nesting in colonies (usually <50 pairs) in tunnels mostly 46–90cm long, dug by both birds into riverbanks, gravel pits and sandy sea cliffs. Builds nest of grass, leaves and feathers at end (unlike bare chamber of Kingfisher, p. 95). Aerial feeder, chiefly over water. Gregarious on migration, roosting in hundreds or thousands in reedbeds. Call note is a harsh twitter, and song is an elaboration of this.

House Martin ■ *Delichon urbicum*
L 12.5cm WS 26–29cm

DESCRIPTION Upperparts deep, metallic blue; flight feathers and tail dull black, these in sharp contrast to pure white underparts and rump. Deeply forked tail. Legs fully feathered white above pink feet. Small black bill.

DISTRIBUTION Summer visitor throughout region except Iceland and high Arctic. Winters across Africa S of Sahara.

HABITS AND HABITAT Colonial nester, mostly in groups of 5 or less. A few birds still nest on primeval site of a cliff with rocky overhang; most now under eaves of a wide variety of buildings, even in city suburbs. Builds a cup-shaped mud nest, lined with grass and feathers, up against overhang, with entrance at top. Migrants roost in trees, and are also believed to roost high in the air then and during breeding season. Song: soft twittering. Call note: hard *chirrrp*.

Swallow ■ *Hirundo rustica*
L 17–19cm WS 32–35cm

DESCRIPTION Long wings, and deeply forked
tail with long, thin outer streamers (2–7cm, male's
longer than female's or juvenile's). Adult: upperparts
and breast-band shiny blue-black, flight feathers dull
black, underparts off-white. Deep red forehead, chin
and throat. When spread, tail shows white spots
near tips of all but outermost feathers. Juvenile:
red is replaced by reddish-buff.
DISTRIBUTION Summer visitor throughout, except
Iceland and high Arctic. Winters in Africa, mostly
S of Equator.
HABITS AND HABITAT All kinds of open country,
often near water, for aerial feeding. Primeval nest
site rare, in entrance of a cave; now mostly on a
ledge in a farm building, shed, stable or church
porch. Gregarious on migration, gathering on
wires and roosting in reedbeds. Call: sharp *witt*
or *witt-witt*. Warbling song, on the wing.

Tawny Pipit
■ *Anthus campestris*
L 17cm WS 25–28cm

DESCRIPTION Long, slim pipit,
like a pale wagtail. Adult: almost
plain sandy brown from crown to
unstreaked rump; underparts white
with buff tint to breast (plus faint
darker streaks) and flanks. White
supercilium and dark loral stripe
in all plumages. Wings at rest are
brown with buff feather edges,
except pale buff fringes to blackish
median coverts, forming a striking
panel. White outer tail feathers. Juvenile: spotted breast and dark-streaked back.
DISTRIBUTION Summer visitor across the region except British Isles and most of
Scandinavia outside far S of Sweden; has declined everywhere. Winters in African
Sahel S of Sahara.
HABITS AND HABITAT Pipits do not hop; they walk. Found on sand-dunes, gravel
pits, and sandy heaths and grasslands. Song in flight: rather monotonous, repeated
che-VEE che-VEE. Call: loud *tseep*.

Water Pipit ■ *Anthus spinoletta*
L 16–17cm WS 23–28cm

DESCRIPTION Summer adult: grey head and nape with bold white supercilium; underparts pale pink from throat, fading to white belly, streaked on flanks; upperparts and wings brown, the latter with blackish centres to tertials and median and greater coverts; the coverts are white-tipped, forming 2 wing bars. Dark tail has white outer feathers. Winter adult: all grey and pink lost; underparts distinctly streaked on white; retains white wing bars and outer tail.

winter   DISTRIBUTION Breeds in mountains at *c.* 600–2,000m above sea-level in S of region (more common further S). Winters in coastal areas.

HABITS AND HABITAT Breeds near watercourses with low plant cover. Winters on lowland wet meadows, estuaries and seashore, including Britain. Call like that of a Rock Pipit (*below*): a sharp *pheest*. Full song, in flight: a rattling tinkling in c. 5 sections, lasting 9–12 seconds.

Rock Pipit ■ *Anthus petrosus*
L 16–17cm WS 23–28cm

DESCRIPTION Predominantly dark, streaked, dirty-looking plumage. Upperparts brownish grey with an olive tint, mantle streaked darker, wing feathers with darker centres; wing bars indistinct buff (cf. Water Pipit, above). Underparts dirty white, tinged buff on breast and flanks, both of these heavily streaked. Outer tail feathers greyish white. Formerly conspecific with Water Pipit, and very like that species in winter. DISTRIBUTION Breeds on coasts of NW France, British Isles, Scandinavia and Finland. N birds disperse to winter on coasts of S British Isles, and from France to Germany and Denmark.

HABITS AND HABITAT Almost always on rocky islands, sea cliffs and rocky seashores. Builds nest in a hollow in cliff or bank by the shore, or under thick vegetation above high-tide line. Call: short, sharp *pseep*. Song in flight like Water's, with fewer but longer phrases.

Meadow Pipit ■ *Anthus pratensis*
L 15cm WS 22–25cm

DESCRIPTION Very hard to tell from Tree Pipit (below) except by voice. Grey-brown above with olive tint, and dark brown stripes on crown and back. Indistinct head pattern. Centres of coverts and tertials blackish, edges pale, wing bars dull. Underparts white with buff at sides, breast and flanks heavily streaked. White outer tail feathers. Yellow-brown legs.

DISTRIBUTION Found throughout region. N birds migrate to winter in British Isles, from Denmark to France, and S to N Africa.

HABITS AND HABITAT Nests in open country – heaths, rough grassland, moors and sand-dunes. The most ground-loving pipit. Deserts uplands in winter for farmland and seashore. Call: weak *ist ist ist*. Song in flight: sequence of weak notes, gathering speed as bird rises and ending in a trill as it 'parachutes' down.

Tree Pipit ■ *Anthus trivialis*
L 15cm WS 25–27cm

DESCRIPTION Plumage differences from Meadow Pipit (above) are slight: yellow-buff breast and flanks, heavily streaked on breast and lightly marked on flanks; more noticeable head pattern with white supercilium and stronger bill; pink legs. Tends to hold itself more erect than Meadow, which creeps along.

DISTRIBUTION Summer visitor to region except N islands, Iceland and Ireland. Winters in tropical Africa.

HABITS AND HABITAT Rough agricultural land, heath, young conifer plantations and birch scrub; avoids total tree cover and completely open ground. Unique among pipits for needing scattered trees for song posts. Song usually starts from a treetop; male then flies up trilling, with repeated notes, and finishes with a musical *SEE-a-SEE-a-SEE-a* as it glides back with upraised wings to the same or a nearby perch. Call: hoarse *teez*.

Red-throated Pipit
■ *Anthus cervinus*
L 15cm WS 25–27cm

DESCRIPTION Summer adult: all upperparts – including rump, tail coverts and crown – rich buff-brown with dark brown streaks, these especially broad on mantle, and dark-centred wing feathers; looks darker than other small pipits. Face, chin, throat and breast rusty red (some birds brighter or paler than others); female usually less warmly coloured. Winter adult: loses red and gains white throat and heavily streaked breast.

DISTRIBUTION Summer visitor to Arctic Fennoscandia and Russia. Winters in E Mediterranean but mostly in E Africa.
HABITS AND HABITAT Breeds above tree-line on shrubby or mossy tundra, and swamps of willow or birch. Winters especially where cattle graze, and around Rift Valley lakes. Birds scarce but widespread in autumn in Poland on their way S, but rare or accidental further W. Migration calls: rasping *teez* or *skee-eaz*, or abrupt *chup*.

Grey Wagtail ■ *Motacilla cinerea*
L 18–19cm WS 25–27cm

DESCRIPTION Has a particularly long tail that constantly wags, so whole rear end rocks. Adult male: upperparts blue-grey, rump yellow-green. White supercilium, white lower border to cheeks, black bib. Underparts yellow, brightest under tail. Wings and tail black, with white wing bar and outer tail. Legs brownish pink, not dark as other wagtails. Adult female: similar but bib white or mottled black, and underparts paler. Immature: white throat, and yellow only under tail (beware confusion with Yellow Wagtail, p. 108).

DISTRIBUTION British Isles, France eastwards to Poland, Czech Republic, S Norway and Sweden. N birds especially migrate to winter in rest of breeding range.
HABITS AND HABITAT Scattered populations in upland districts on swift-flowing, rocky, mostly shallow rivers and streams with wooded banks. In winter, also by lowland lakes and streams. Call: distinctive metallic *tzi-tzi*.

Pied Wagtail

■ *Motacilla alba yarrellii*
L 18cm WS 25–30cm

DESCRIPTION Male: black crown, nape and upperparts. Black throat and chest, white face and rest of underparts. White wing bars, white edges of tertials, and white outer feathers to black tail. Female: dark grey upperparts (including nape) and wing coverts, and black rump. Both have well-patterned pied wings. The very similar **White Wagtail** *Motacilla alba alba* is more common on the Continent. Summer male resembles male Pied but with pale grey upperparts; summer female has black crown and dull grey upperparts; winter adult has plain grey from crown to rump, grey ear coverts, and black gorget on breast.

DISTRIBUTION Pied breeds in British Isles and locally on nearby Continental coast; in non-breeding season, birds disperse southwards or are short-distance migrants. White is found everywhere else, including Iceland and, rarely, British Isles; migrates to winter in W Europe and Mediterranean basin, e.g. many Icelandic birds head to Britain.

TOP: *Pied Wagtail;* ABOVE: *White Wagtail*

HABITS AND HABITAT Both favour open country, especially around farms and cultivation with nearby water. Ground feeders on insects, walking with nodding head and wagging tail. Territorial breeders, but social in autumn and winter, with roosts of hundreds in reedbeds, on trees, or on buildings in towns. Flight call: *tschizzick*, Pied's said to be harsher than White's. Song: lively twittering or warbling.

Yellow Wagtail

Yellow Wagtail
■ *Motacilla flava* L 17cm WS 23–27cm

DESCRIPTION 3 clearly identifiable subspecies breed in region. (1) **Blue-headed Wagtail** M. f. *flava*: male has bluish-grey crown and darker ear coverts, and white supercilium; back and rump yellowish green; wing coverts black with pale greenish yellow fringes and tips (so 2 wing bars); tail black with white outer feathers; whole of underparts bright yellow. (2) M. f. *thunbergi*: male has dark slate-grey crown and nape, and almost black cheeks, with or without whole or faint white supercilium. (3) **Yellow Wagtail** M. f. *flavissima*: male very distinct, with yellow forehead, yellowy-green crown and cheeks, yellow supercilium, and upperparts washed with yellow. All females are duller above and less yellow below, with a 'dirty' supercilium; form best identified when with a male.

DISTRIBUTION *Flava* wagtails are found across Europe and Asia. About a dozen forms, largely differentiated by head colour, are named. Our regulars are summer visitors, wintering in tropical Africa: (1) is a W and central European race; (2) breeds in N Fennoscandia; (3) breeds in British Isles and coastal France. Other subspecies known only by scientific names from the S and E occur on passage, bringing further confusion to identification!

HABITS AND HABITAT (1) and (3) haunt lowland water meadows, riversides and fens with low, dense vegetation. (2) N birds inhabit sphagnum and peat bogs with shrubs or stunted trees. All nest on the ground in territories, and feed on a huge variety of invertebrates. Gregarious on migration. Often found hunting insects around the feet of cattle. Most often spotted when a passing bird calls a shrill, rather plaintive *tsweep*.

Blue-headed Wagtail

Wren ■ *Troglodytes troglodytes*
L 9–10cm WS 13–17cm

DESCRIPTION Unmistakable. Sexes and ages alike. Looks tiny, especially owing to its habit of holding its tail vertical. Reddish brown above, whitish brown below, all with fine, darker barring. Pointed, quite long bill.
DISTRIBUTION Iceland, British Isles and across region except N Fennoscandia. Mainly resident; most N and E populations migrate S and W.
HABITS AND HABITAT Dense undergrowth in a wide range of places: woodland, islands, gardens, hedgerows, and moorland with heather and Bracken *Pteridium aquilinum*. Active, hunting invertebrates in low cover. Rapid, straight, whirring flight between bushes. Builds a completely domed nest. Call: rattling alarm, and hard *tic tic tic*. Song: notes and trills for 4–6 seconds, amazingly loud for a small bird.

Dipper ■ *Cinclus cinclus*
L 18cm WS 25–30cm

DESCRIPTION Adult: head and nape chocolate-brown; rest of upperparts more slatey with black feather edges. Underparts white from chin to breast, the rest blackish brown (birds in British Isles have a chestnut band between the white and brown). Noticeable white eyelid when bird blinks. Juvenile: mottled grey. 1st-year: like adults but with white tips to wing coverts.
DISTRIBUTION Resident N and W British Isles; some in suitable habitats in France and to Scandinavia eastwards.
HABITS AND HABITAT Mostly on fast-flowing streams and rivers with rocks and boulders in hilly regions. Territorial. Flight low and direct along river. Bobs on hinged legs on boulder. Feeds on aquatic insects and larvae, often underwater, walking along the bottom, or swims with flicking wings. Song: rippling warble. Call: short, sharp *zit zit* as it flies.

Waxwing

■ *Bombycilla garrulus*
L 18cm WS 32–35cm

DESCRIPTION Unmistakable. Crest, head, back and wing coverts vinaceous brown; rump grey, tail black with yellow tip; narrow black eye-stripe with chestnut above and below; black bib; reddish vent. Amazing wing pattern: black or blackish flight feathers; white-tipped primary coverts; white-tipped secondaries with wax-red tip beyond that; primaries end with a white and yellow 'V'.
DISTRIBUTION Breeds in Arctic Sweden, Finland and Russia eastwards. Regularly winters in E and N Europe.
HABITS AND HABITAT Breeds in lowland and upland forest, preferably conifers. Feeds on insects in the breeding season by flycatching from treetops; changes to fruits as winter comes. If fruit crop fails, large flocks erupt to find Rowan *Sorbus aucuparia*, Common Hawthorn *Crataegus monogyna*, cotoneaster, etc., as far W as France and Ireland.

Dunnock ■ *Prunella modularis*
L 15cm WS 19–21cm

DESCRIPTION Slender-billed bird. Warm brown above, and lead-grey head and below, recalling female House Sparrow (p. 144; hence pre-1950s English name Hedge Sparrow). Brown upperparts streaked with blackish brown; sides of breast and flanks buff-brown, streaked darker; brownish crown and ear coverts.
DISTRIBUTION Throughout region except Iceland and high Arctic. N and Continental populations migrate S and W; resident in British Isles and W France, or disperses short distances.
HABITS AND HABITAT Favours mixed woodland along river valleys, hedgerows, spinneys, gardens and low scrub on moorland. Unobtrusive ground feeder, creeping along, belly near ground. Call: loud, shrill *tseep*, and short, loud, high-pitched warble.

Nightingale ■ *Luscinia megarhynchos*
L 17cm WS 23–26cm

DESCRIPTION Sexes alike. Uniform brown above with plain rusty-red tail and rump. Pale greyish-brown underparts with paler throat and belly. Noticeable black eye with whitish eye-ring.

DISTRIBUTION Summer visitor to SE England, and from France to Poland. Winters in Africa between Sahara and forest.

HABITS AND HABITAT Mostly solitary with traditional territories, in thickets and woods with rich undergrowth, near water; especially fond of coppiced woods. Feeds (on invertebrates) and nests on the ground. Skulking, so hard to see, but much loved in literature and folklore because of its splendid song, a very varied, melodious succession of phrases, whistles and trills, with a recurring group of rapid *chooc chooc chooc* notes, and higher-pitched, flute-like, slow *pioo pioo pioo* in a rising crescendo. Sings by day too.

Thrush Nightingale
■ *Luscinia luscinia* L 17cm WS 24–26cm

DESCRIPTION Very similar to Nightingale (above), but upperparts more drab and with less contrast between mantle and rump and tail; underparts greyish buff with pale throat, the rest faintly mottled and barred with brown; some birds have a clear brown breast-band.

DISTRIBUTION The 'Nightingale' of E and N Europe, with a slight overlap with its relative: summer visitor to Germany, Denmark and S Fennoscandia eastwards. Winters in E Africa.

HABITS AND HABITAT Breeds in thick cover in river valleys, alder carr, and Hazel *Corylus avellana*, birch and willow scrub. Feeds mostly on ground, and nests among thick roots or leaf litter. Song: very varied and loud song, like Nightingale's but recognisable with practice – almost always ends in a rattle, and more often includes the introductory *pioo* crescendo.

Bluethroat ■ *Luscinia svecica*
L 14cm WS 20–23cm

DESCRIPTION Adult male: unmistakable. Upperparts and wings dark brown, supercilium whitish; throat and upper breast blue, bordered below with black–white–chestnut bands; rest of underparts off-white; diagnostic bright chestnut base to black tail shows well in flight. Adult female: lacks or has reduced amount of blue, instead having blackish moustache joined to broken necklace. 1st-winter: similar to adult female. N birds have a red spot in centre of breast; French and central European birds have a white spot.

DISTRIBUTION Scattered in W and central areas, more in Fennoscandia. Summer visitor, wintering in sub-Saharan Africa N of forests. Scarce passage migrant to E coast of Britain.

HABITS AND HABITAT Secretive and mostly ground-loving, in shrubby wetlands and overgrown riversides with alder and willow, from sea-level to high fjelds (plateaux). Song: loud and tinkling, with *ting-ting-ting* phrase. Call: *tacc tacc*.

Robin ■ *Erithacus rubecula*
L 14cm WS 20–22cm

DESCRIPTION Adult: sexes alike. Upperparts, wings and tail olive-brown. Forehead, cheeks and breast orange, separated from brown by band of blue-grey. Flanks warm buff, rest of underparts white. Juvenile: quite different. Brown above, buff below, with no orange on breast, but buff spots all over the brown, and small, darker spots on the breast.

DISTRIBUTION Throughout region except Iceland and far N of Fennoscandia. Largely resident in British Isles; birds from the N and E winter in the W, and S to N Africa.

HABITS AND HABITAT Prefers well-shaded woods and copses with some open spaces on which to feed, particularly ground turned by animals or humans. In British Isles and the W especially, a regular in parks, gardens and farmland. Call: scolding *tic tic* and high-pitched *tswee*. Song: melodious, rather sad-sounding warble.

Redstart
■ *Phoenicurus phoenicurus*
L 14cm WS 21–24cm

DESCRIPTION All ages and both sexes have orange-chestnut rump and tail (with dark centre), which is constantly quivered up and down. Adult male: forehead white; rest of upperparts blue-grey, including wing coverts; wings blackish brown; throat and upper breast black, sharply divided from orange underparts. Adult female and immature: grey-brown upperparts and pale orange-buff underparts, fading to whitish belly.
DISTRIBUTION Summer visitor to everywhere except Iceland, Ireland, Northern Isles and high Arctic. Winters in tropical Africa N of Equator.
HABITS AND HABITAT Breeds in old deciduous woodland and hill country with scattered trees and stone walls. Call: plaintive *wheet*, often linked with *tooick*. Brief, melodious song.

Black Redstart
■ *Phoenicurus ochruros*
L 15cm WS 23–26cm

DESCRIPTION All ages and both sexes have a dark-centred orange tail like Redstart (above). Adult male: dark grey above with dark underparts, these blackest on face and greyest on belly; orange-buff under tail. Wings brownish black with off-white panel on inner secondaries. Female and immature: mouse-grey, and tail-end colour less bright.
DISTRIBUTION Summer visitor to SE England and France eastwards to about 60°N and 40°E. Winters in W Europe, including S Britain and Mediterranean basin.
HABITS AND HABITAT Breeds on stony, craggy hillsides, regularly by farmsteads. Often winters on rocky coasts. Has extended its distribution into towns where waste ground and buildings simulate stony ground and cliffs. Call: quiet *tsip*, often preceding alarm *tuc tuc*. Song: quick warble with a quieter, strange sound like crackling cellophane towards the end.

Wheatear ■ *Oenanthe oenanthe*
L 14.5–15.5cm WS 26–32cm

DESCRIPTION Male: body plumage pattern is diagnostic, and white tail tipped by a broad black upside-down 'T' is distinctive. Female: similarly patterned, but wings, crown, cheek-patch and back all brown-toned, underparts usually buffer.

DISTRIBUTION Found from Arctic southwards, from sea-level to >3,000m. Summer visitor, wintering in Africa at 0–20°N. Often the first songbird to arrive in the region, commonly in Mar. Autumn migration is protracted from Aug to Oct.

HABITS AND HABITAT Breeds from Arctic tundra southwards, and on sand-dunes, cliff tops, moors and mountains. Nests in a hole in a rock or wall, or in a Rabbit burrow. Eats mainly insects and other invertebrates, and berries in autumn. Wary, and if pressed dashes away showing its striking tail pattern. Call: *chak*. Alarm: *weet-chak, chak*. Song: energetic, short, pleasing warble.

Whinchat ■ *Saxicola rubetra*
L 12.5cm WS 21–24cm

DESCRIPTION Adult male: distinctive dark brown head with a broad, long white supercilium. Dark brown upperparts, orange breast, white belly. Tail is black with white sides at base, noticeable in flight. Adult female: similarly patterned but duller. Juvenile: like a dull female.

DISTRIBUTION Summer visitor to region, mostly further E and N than the Stonechat (p. 115), into the far N of Fennoscandia. Winters in Africa S of Sahara.

HABITS AND HABITAT Favours open meadows and scrub, and hilly, Bracken-covered slopes, where it nests on the ground. Feeds mainly on invertebrates. British numbers are down 26% since the 1990s, probably through loss of habitat in Africa. Call: harsh *tzee*. Song: long burst of phrases, some musical, some scratchy, some imitating other species.

Stonechat ■ *Saxicola torquata*
L 12.5cm WS 18–21cm

DESCRIPTION Adult male: distinctive black head
and throat, bordered by white. Back dark brown,
white rump streaked brown. Wings dark brown with
a variable-sized white panel. Breast orange, shading
to white on belly. Adult female: mottled brown head
and upperparts, and less bright underparts. Juvenile:
like a heavily spotted female.
DISTRIBUTION Found in suitable habitats in the
British Isles, especially in the W, and from France to
Germany and Ukraine.
HABITS AND HABITAT Mostly insectivorous and
resident, but the most N and E populations move S
in winter. Resident populations suffer in cold spells.
Prefers cover like gorse and other scrub, heath,
sand-dunes and young plantations. Nests on or near
ground. Shuns extensive grasslands and intensive
agriculture. Call: *tchak*, like knocking 2 stones
together. Song consists of short, scratchy phrases.

Mistle Thrush
■ *Turdus viscivorus*
L 27cm WS 42–47.5cm

DESCRIPTION Adult: sexes similar.
Whitish underparts covered with
large, wedge-shaped black spots;
flanks and breast marked with
buff. Upperparts and wings greyish
brown with conspicuous greyish-
white fringes. Tail grey-brown
with diagnostic white tips to outer
feathers. White underwing striking
in flight. Juvenile: spotted white on
head, mantle and wing coverts.
DISTRIBUTION Breeds across
region except Iceland, but much less commonly in Norway. N and E populations move
S and W in winter into milder parts of the breeding range, forming small flocks.
HABITS AND HABITAT Found in orchards, woods, farmland, parks and gardens,
vigorously defending breeding territory. Eats invertebrates and, in autumn and winter,
berries. Call: harsh, distinctive rattle. Song: short, loud, flutey phrases, which are
far-carrying, even in wild weather.

Redwing ■ *Turdus iliacus*
L 21cm WS 33–34.5cm

DESCRIPTION Sexes alike. Long, creamy-white supercilium contrasts with brown head. All upperparts dark, warm brown, darker on flight feathers. Breast yellowish buff on sides, with dark brown streaks. Underparts white, streaked with lighter brown. Flanks and underwing chestnut-red (cf. Song Thrush's buff underwing, below).
DISTRIBUTION Breeds in Iceland, across Scandinavia, around the Baltic and into Siberia. Colonised Scotland in 1925. British population <100 pairs compared with *c.* 1 million in Sweden. Winters in W and S Europe.
HABITS AND HABITAT Nests in open woods, thickets and scrub, but winters on grassland, stubble, in root crops and in open woodland. Eats a wide variety of invertebrates, and berries in autumn and winter. In winter, roosts are large flocks. As night migrants, their *see-ip* calls keep the flock together and alert observers below. Song is variable: 4–6 flutey notes plus warbling.

Song Thrush ■ *Turdus philomelos*
L 23cm WS 33–36cm

DESCRIPTION Brown above, white below with buff tint to breast and flanks; underparts streaked with black-brown arrow-shaped spots (streaked, not spotted like Mistle Thrush, p. 115). Orange-buff underwing, visible in flight.
DISTRIBUTION The most common and widespread thrush right across the region except the far N and Iceland. All populations move S or W in winter; British birds head S and are replaced by N and E birds.
HABITS AND HABITAT Woodland, parks and gardens with plenty of shrubs. Its nest of grass and moss is uniquely lined with mud and decayed wood. Eats invertebrates, fruit and snails, which are broken on a stone ('anvil'). Call: *sipp*. Alarm: rapid, repeated, scolding *tchuk-tchuk-tchuk*. Song: loud, sustained, characteristic series of phrases, each often repeated 2–4 times.

Blackbird ▪ *Turdus merula* L 24–25cm WS 34–38.5cm

DESCRIPTION Adult male: the only all-black European bird with a bright golden-yellow bill and long tail. Orange-yellow eye-ring. Immature male: bill dark brownish horn, and plumage dull black. Female: is a 'brownbird' – all dark brown, underparts often with a rufous tone and thrush-like mottling on breast. Juvenile of both sexes before autumn moult: as female but more rufous, and more spotted below.

DISTRIBUTION Breeds throughout, but only sparingly N of 65°N. N populations are migrants, moving to S and W of breeding range; others are resident.

HABITS AND HABITAT Breeds in most places where trees are present, but also on moors. Eats insects, earthworms and, in autumn and winter, wild fruits. Hunts in leaf litter. Call: *see.* Alarm: shrill chatter. Song: variety of flute-like, musical phrases.

ABOVE: *male;* RIGHT: *female*

Ring Ouzel ▪ *Turdus torquatus* L 23–24cm WS 38–42cm

DESCRIPTION Adult male: sooty black with a broad white crescent across breast; greyish edges of secondaries and wing coverts show as a grey patch in flight and at rest. Adult female: narrower, brown-tinged gorget; duller plumage with brownish-grey feather edges, which create a scaly effect. Juvenile: whitish chin, no breast-band and more noticeable pale feather edges. SE birds have white-scaled underparts.

DISTRIBUTION Summer visitor to N and W British Isles, Scandinavia, Finland, S Germany and Poland. Winters in Iberia and N Africa.

HABITS AND HABITAT Nests on mountainsides and moorlands, with open ground for feeding on insects and worms. Nest is on or near ground in low vegetation; or in Scandinavia, often in a tree. Call: loud *tac-tac-tac.* Song: piping, sad-sounding *pee-pee-pee.*

Fieldfare ■ *Turdus pilaris*
L 25.5cm WS 39–42cm

DESCRIPTION Unmistakable. Slate-grey head, nape and rump contrast with chestnut back, black tail and white underwing. Throat and breast golden brown, streaked black. Rest of underparts white, flanks streaked black.
DISTRIBUTION Breeds widely in Scandinavia, Finland and Germany, mostly in wooded country. A few breed elsewhere in region. N and E populations winter to W and S.
HABITS AND HABITAT Breeds in open woodland, scrub, gardens and parks. Gregarious. Feeds on a variety of invertebrates, and fruits in autumn and winter. Many hundreds roost together in evergreens in winter; most commonly breeds in colonies of up to 40–50 pairs. All feeding is done on neutral ground. A noisy, aggressive bird at nest sites and in defence of winter food sources. Call: *tchak tchak*; readily identifies migrating flocks. Song: a weak warble with some wheezes and chuckles.

Barred Warbler ■ *Sylvia nisoria*
L 15.5cm WS 23–27cm

DESCRIPTION Adult male: distinctive, with grey head, face and upperparts, and pale yellow eye. Wings dark brown-grey with 2 pale bars. Underparts dull white with dark grey-brown crescents, emphasised by white tips. Tail long and broad, dark brown with white edges. Female: duller and browner above, and barring less clear below. Juvenile: grey-brown, unbarred, dark-eyed.
DISTRIBUTION Summer visitor, from about 10°E in Germany, eastwards to the Baltic and on into Asia. Winters chiefly in Kenya. Rare autumn visitor W of its breeding range.
HABITS AND HABITAT Lowland bird of thick cover and riverine woodland. Insectivorous, also eating fruits and berries in autumn. Males display conspicuously, fanning their tails, then perform song flight for 5–10 seconds, often preceded by wing-clapping. Call: harsh *charr*. Song: short, rich warble with harsh call notes mixed in.

Garden Warbler

■ *Sylvia borin* L 14cm WS 20–24.5cm

DESCRIPTION Adult: sexes alike. Rounded head, stubby bill, pale eye-ring, plain brown plumage above (darker on wings), and pale buff below, fading to white on belly and under-tail coverts. Juvenile: more strongly marked buff below and tawny above. DISTRIBUTION Summer visitor. Widely distributed, except Iceland and N Fennoscandia. W populations winter in W Africa, E populations in E and S Africa. HABITS AND HABITAT Rare in gardens! Breeds in broadleaf woodland with thick undergrowth, scrub and young conifers. Often sings from thick cover, so is hard to see. Eats mostly insects, plus some fruit and berries. Avoids competition with Blackcap (below) by arriving later and feeding lower down. Song can be confused with Blackcap's but is pitched lower, with longer phrases.

Blackcap

■ *Sylvia atricapilla*
L 13cm WS 20–23cm

DESCRIPTION Beware confusion with Marsh and Willow tits (p. 133). Adult male: black forehead and crown, ash-grey nape and face. Upperparts ashy brown, darker on tail and primaries. Chin, breast and flanks grey; belly and under-tail coverts white. Adult female: similar but cap bright red-brown and upperparts browner. Juvenile: like a dull female. DISTRIBUTION Summer

male

female

visitor from S Scandinavia to much of region (not Iceland). W European birds usually migrate SW to W Africa, but some winter in British Isles; the rest fly SE to E Africa. HABITS AND HABITAT Forages and sings in treetops in open woodland, copses with thick undergrowth and parks. May be found wintering in parks and gardens in town. Eats insects and, especially in autumn, fruit and berries. Call: loud, repeated *tac*. Song: loud, rich warble, rising in pitch.

Lesser Whitethroat

■ *Sylvia curruca* L 12.5–13.5cm WS 16.5–20.5cm

DESCRIPTION Looks less like a whitethroat than the Whitethroat (below). Adult: Sexes similar. Grey above and white below. Dark face mask against slate-grey head. Primaries and tail browner and darker than rest of upperparts. White outer tail feathers. Juvenile: like female but rather browner.

DISTRIBUTION Widespread from England and NE France, eastwards into Asia and northwards to S Fennoscandia. Summer visitor, wintering mostly in NE Africa.

HABITS AND HABITAT Active but skulking in woodland edges, thick hedges and shrubberies. Mainly insectivorous, plus summer fruits and berries. Call: hard *tack* or *churr*. Song: unlike any other warbler's – a loud rattle and a quiet warble. The rattle may carry 200m and sound like the whole song, and is often preceded by a low warble, audible from only a few metres away.

Whitethroat

■ *Sylvia communis*
L 14cm WS 18–23cm

DESCRIPTION Adult male: grey cap to below eye, whitish eye-ring, pure white chin and throat; breast pale pinkish buff, rest of underparts white; upperparts dull brown. Noticeable rufous wing panel. Dark brown tail with white outer feathers. Female and juvenile: same tail and wing patterns as adult male but plumage is duller.

DISTRIBUTION Summer visitor, breeding widely across Europe from 65°N in Fennoscandia southwards (but not central region, or Iceland). Winters mostly in tropical Africa S of Sahara.

HABITS AND HABITAT Found in most open habitats with thickets and shrubs. Population crash in 1968–69; numbers have since recovered slowly. Cock builds several nests; female may complete one. Call: scolding *charr* and sharp *tac*. Song: short, rapid, chattery warble, often in a dancing song flight.

Dartford Warbler

■ *Sylvia undata*
L 12.5cm WS 13–18.5cm

DESCRIPTION Small-bodied
warbler with a long tail. Adult
male: dark slatey brown above
with greyer head, and wine-red
below. Wings brownish black.
Tail grey-black with narrow white
edges. Eye and eye-ring orange to
red. Female and juvenile: browner
on upperparts and paler below. In
poor light, all look uniformly dark.
DISTRIBUTION Only in S
England and W France in our
region, the extreme edge of its
normal S European range. Mostly resident, so hard hit by cold winters.
HABITS AND HABITAT Likes low, dense cover, on coastal scrub and lowland heather
or gorse. Conservation of its habitat is essential if we are to keep this bird in our region.
A skulking bird, feeding exclusively on invertebrates. Call: *tuc*. Alarm: grating *tchirr*.
Song: a musical chatter with some liquid notes.

Zitting Cisticola

■ *Cisticola cisticola* L 10cm WS 12–14.5cm

DESCRIPTION Smallest warbler in region
aside from Goldcrest and Firecrest (p. 128);
until recently known as Fan-tailed Warbler.
Sexes similar. Buff above and below, with
strong blackish-brown streaks on crown,
mantle and wings. Short brown fan-shaped
tail is tipped black then white, noticeable
in flight.
DISTRIBUTION Coastal France to the
Netherlands. Resident. Essentially a
Mediterranean species in Europe that spread
in the 19th century.
HABITS AND HABITAT Almost always in
grasslands or other vegetation no higher than
1m. Call: loud *tsip* or *zit*. Song: diagnostic, in
an undulating flight, a sustained *zit zit zit zit…*,
with each note at the top of the rise – hence
its common name.

Cetti's Warbler ■ *Cettia cetti*
L 13.5cm WS 15–19cm

DESCRIPTION Has short, rounded wings and
a long, broad, graduated tail. Sexes similar.
Upperparts and wings uniform chestnut-brown.
Off-white eye-ring and short grey supercilium.
Chin and central underparts off-white, the rest
grey-brown, darker on flanks and under-tail
coverts, which are tipped white.
DISTRIBUTION Saw a marked spread in the
20th century from its Mediterranean base,
reaching N France and the Low Countries
in the 1960s and first breeding in Britain in
1972. Resident, so suffers in cold winters.
HABITS AND HABITAT A skulking bird of bushy
places by watersides, swamps and marshes. Males
often mate with 2–4 females. Very distinctive,
diagnostic song is a sudden explosion of sound
lasting a few seconds.

Savi's Warbler ■ *Locustella luscinioides*
L 14cm WS 18–21cm

DESCRIPTION Similar to other plain warblers,
but note the dark, unstreaked reddish-brown
head and upperparts, and faint supercilium.
Underparts brownish white with rufous brown
along sides of breast to under-tail coverts. Wings
and tail reddish brown. Tail broad and graduated.
DISTRIBUTION Very fragmented distribution
across region to Poland and the Baltic, and
thence eastwards nearly to the Urals. Summer
visitor, wintering in Africa S of Sahara and N
of forest zone. A few pairs breed in Britain.
HABITS AND HABITAT Unbroken reedy
swamps and fens, overgrown fringes of lakes.
The species is in danger wherever its wetlands
are threatened by development. Eats adult and
larval flies, bugs, spiders and beetles. Call: *pit*;
alarm rattle sharper. Song: accelerating, reeling,
ticking sound, remarkably like that of a Roesel's
Bush-cricket *Metrioptera roeselii*! Often sings
through night.

Grasshopper Warbler
■ *Locustella naevia* L 12.5cm WS 15–19cm

DESCRIPTION Sexes alike. Upperparts olive-brown, spotted and streaked with dark brown. Wings darker, with buff to reddish fringes. Tail reddish brown, barred darker. Underparts mostly buff, but almost white on chin, throat and centre of breast and belly. Under-tail coverts streaked with brown. DISTRIBUTION Thinly distributed eastwards to around the Baltic, but not Iceland. Summer visitor, believed to winter in W Africa S of Sahara. HABITS AND HABITAT Skulks in marshes, thick hedges, heathland and new plantations. Song, heard mainly at dusk or dawn, is strange, uniformly pitched and sustained, like trilling insects; it is far-carrying, even up to 1km, but is inaudible to some people. Sings seemingly without pause, but actually with very short breaks – one 'uninterrupted' song lasted 110 minutes!

Sedge Warbler
■ *Acrocephalus schoenobaenus*
L 13cm WS 17–21cm

DESCRIPTION Adult: sexes similar. Olive-brown upperparts with dark streaks. Tawny unstreaked rump, dark brown tail. Wings buff-brown with lighter edges to tertials and greater coverts. Underparts off-white, whitest on throat and belly, more rufous on flanks. Streaked crown, long creamy-white supercilium. Juvenile: creamier supercilium, yellower underparts and distinct brown-spotted breast. DISTRIBUTION Summer visitor, from high Arctic southwards. Absent from mountainous Scandinavia and Iceland. Winters in Africa S of Sahara. HABITS AND HABITAT Found in low, lush vegetation in moist habitats. Birds gain weight rapidly on S migration at stopovers – some may double their weight and be able to fly non-stop from S Britain to S of Sahara. Call: *tuc*. Alarm: *churr*. Song: loud, varied mix of harsh and musical notes.

Reed Warbler

■ *Acrocephalus scirpaceus*
L 13cm WS 17–21cm

DESCRIPTION Sexes similar. Upperparts uniform olive-brown, with a more rufous rump and upper-tail coverts, and darker brown wing-tips. Underparts white with buff under-tail coverts and sides of breast. Indistinct cream supercilium.
DISTRIBUTION S Britain eastwards to around the Baltic, where it is found mostly S of 60°N. Summer visitor, wintering in Africa S of Sahara as far as Zambia.
HABITS AND HABITAT Insectivorous; closely associated with reedbeds (a vulnerable habitat, so managed conservation of this is necessary to preserve the species). Will also breed in willowherb and recently reported in crops of Rape *Brassica napus*. Builds a unique nest woven round several stems. Call: *churr-churr*; alarm harsher. Song: low, guttural churring, with long phrases. Along with Meadow Pipit (p. 105) and Dunnock (p. 110), is one of the main hosts of the Cuckoo (p. 89).

Marsh Warbler

■ *Acrocephalus palustris*
L 13cm WS 18–21cm

DESCRIPTION Very hard to separate from Reed Warbler (above). Note more olive-brown upperparts, lack of rufous rump, shorter bill, more round-headed and pot-bellied appearance, long wings, pale legs and astonishing song.
DISTRIBUTION Breeds at 45–60°N from E France and the Low Countries eastwards across Europe. Britain is on very edge of its range, with fewer than 20 pairs, yet as many as 70,000 pairs nest in the Netherlands. Summer visitor, wintering in E Africa, spending 3 times as long there as it does in Europe.
HABITAT Dense, low, rank vegetation. Eats invertebrates and berries. Sings in both Africa and Europe. Call: *tchuc*. Alarm: *chirrr*. Song: prolonged and very musical, with a remarkable amount of mimicry; an average repertoire may contain 45 African and 31 European songs.

Great Reed Warbler
▪ *Acrocephalus arundinaceus*
L 19–20cm WS 25–29cm

DESCRIPTION Like a large Reed Warbler (p. 124) with a stouter, dark-tipped bill. Male: upperparts warm olive-brown, crown darker, rump more fawn and wing coverts edged rufous. Creamy supercilium, dusky eye-stripe, pale cream eye-ring. Underparts creamy buff, but more buff on flanks, and off-white chin and throat. Female: tends to be brighter above and less white below. DISTRIBUTION Breeds widely in restricted habitats, but not Britain or Iceland, and only S Sweden and Finland, in isolated populations. Summer visitor, wintering in Africa S of Sahara. Rare visitor to Britain.
HABITS AND HABITAT Territorial, especially in dense *Phragmites* reedbeds. Many males are polygynous, with as many as 3 females. Call: *chak*. Alarm: harsh chatter. Song: very loud, harsh churrs and rattles, with remarkable variety and volume, carrying as far as 1km.

Icterine Warbler
▪ *Hippolais icterina* L 13.5cm WS 21–24cm

DESCRIPTION Adult: basically green above and yellow below, with a long bill, rather flat crown and yellow face; long wings, reaching at least to upper-tail coverts. Wings have a distinct pale panel. Legs bright blue-grey. Juvenile and late-summer adult in worn plumage: browner above and whiter below. DISTRIBUTION Breeds from NE France eastwards, and N to 60–65°N in S Fennoscandia and Russia. Summer visitor, wintering in Africa S of Equator. Rare annual visitor to British Isles. HABITS AND HABITAT Sunny, wooded lowlands, cultivated lands and gardens. Eats chiefly adult and larval insects, plus various fruits. The cup-shaped nest of grasses, moss and bark, lined with hair, feathers and finer grasses is much more substantial than that of most other warblers. Call: *tec* and, in spring, a diagnostic *deeteroo*. Song: loud, remarkably long warble, which may last for up to 40 seconds and carry 500m.

Willow Warbler

■ *Phylloscopus trochilus*
L 10.5–11.5cm WS 17–22cm

DESCRIPTION *Phylloscopus* warblers are popularly known as leaf warblers for their colours and feeding habitats. Adult: olive-green above, yellowish white below, cleaner colours than similar Chiffchaff (below). Pale yellow supercilium, orange-brown legs (cf. Chiffchaff's). In summer become browner above and whiter below from feather abrasion. Autumn juvenile: much yellower supercilium, throat and breast.
DISTRIBUTION Commonest leaf warbler in region. Breeds far into the Arctic in scrub beyond the tree-line, but not Iceland. Winters in tropical Africa beyond the Sahara.
HABITS AND HABITAT Woods and forests, but mainly coppices and scrub. Active little bird, searching canopy for insects. Builds well-concealed nest on ground. Call: plaintive, disyllabic *hoo-eet*. Song: lovely cascade of pure notes, dying away towards the end and lasting *c*.3 seconds.

Chiffchaff ■ *Phylloscopus collybita*
L 10–11cm WS 15–21cm

DESCRIPTION Like a less streamlined Willow Warbler (above), with less yellow tint. Dull brownish olive above; dull, pale yellow below, shading to buff flanks. E forms are noticeably greyer above and whiter below. Pale yellow supercilium, pale eye-ring, contrasting dark eye and dark legs (cf. Willow Warbler's).
DISTRIBUTION 2nd commonest leaf warbler in region, breeding throughout except Iceland and far N.
HABITS AND HABITAT Early-arriving summer visitor to open woodland with a good shrub layer, copses and hedgerows. Most Chiffchaffs winter around Mediterranean and across Africa S of Sahara. 500–1,000 winter in British Isles; they remain insectivorous, so cold winters kill many. Call: monosyllabic *hweet, hweet*. Song: diagnostic *chiff-chaff-chiff-chaff...*, with more variation than name suggests.

Wood Warbler

■ *Phylloscopus sibilatrix*
L 12cm WS 19–24cm

DESCRIPTION Looks 3-coloured: yellowish-green upperparts; bright yellow supercilium, throat and breast; pure white belly and under-tail coverts. Tail and wings, however, are brown.

DISTRIBUTION Summer visitor to Britain (rarely Ireland), eastwards to region's limit except far N. Winters in Africa, on or about Equator, after an unbroken flight across Sahara.

HABITS AND HABITAT Prefers hilly terrain, and woodland with good canopy (Beech *Fagus sylvatica* and oaks especially). Forages for insects in canopy; hovers well. Builds a domed nest on ground. Usual call: plaintive *pew, pew.* Song: initially a repeated note, starting slowly but going into a passionate trill, then a series of notes like the call.

Greenish Warbler

■ *Phylloscopus trochiloides*
L 10cm WS 15–21cm

DESCRIPTION Very difficult to identify quickly because it is so like other leaf warblers. Carefully observe head pattern, clear wing bar, and call and song. Pale greyish olive above, dull white below. Long yellowish-white supercilium reaching nearly to nape, dark eye-stripe, pale wing bar (in fresh plumage). Legs a variable shade of brown.

DISTRIBUTION Small populations in N Germany and S Sweden, then more widely in S Finland, Lithuania and eastwards into Asia. Summer visitor, wintering on the Indian sub-continent from Himalayan foothills southwards. Vagrant to Britain in autumn.

HABITS AND HABITAT Found in open woodland, copses and overgrown orchards. Active in the canopy, hovering and flycatching. Call: distinctive *chee-wee*, 2nd note lower. Song: medley, like that of Wren (p. 109).

Goldcrest ■ *Regulus regulus*
L 9cm WS 13.5–15.5cm

DESCRIPTION Adult: tiny (only 4.5–7.5g), with dull greenish upperparts, 2 white wing bars and pale olive-green underparts, darker on flanks. Eyes appear large and black on whitish face; crown on both sexes yellow, lined with black; displaying male raises crest to reveal orange centre. Juvenile: lacks crest.
DISTRIBUTION Widespread resident, except Iceland; migrates S and W in winter, from only the harshest weather.
HABITS AND HABITAT Usually breeds in well-grown conifers, but will inhabit deciduous woods, gardens, cemeteries and parks if there are suitable conifers. Insectivorous year-round, so suffers severely in bad winters. Censuses have shown that populations can recover well. Calls and song are too high-pitched for some people to hear. Tiny size, restlessness and high-pitched voice often make it difficult to spot.

Firecrest ■ *Regulus ignicapillus*
L 9cm WS 13–16cm

DESCRIPTION Tiny. Diagnostic patterned head: golden crown lined each side with black, and broad white supercilium underlined by black eye-stripe; displaying male raises crown to reveal orange 'fire crest'; female's crown is yellow. Both sexes have a bronze-coloured patch on sides of neck. Upperparts bright olive-green, and 2 white wing bars; tail is darker and browner; all underparts are white.
DISTRIBUTION Mostly found from France to Poland. N and E populations winter in Mediterranean basin and far W. Breeding range has expanded W. First bred in England in 1962.
HABITS AND HABITAT Spends more time in broadleaved trees and shrubs than the Goldcrest (above), and in conifers feeds in less dense branches; also prefers larger prey – arthropods, aphids, spiders and insect larvae. Call: very high-pitched *zit zit*, lower than Goldcrest's. Song: rapid string of calls.

Spotted Flycatcher

■ *Muscicapa striata* L 14.5cm WS 23–25.5cm

DESCRIPTION Adult: upperparts, wings and tail plain grey-brown. Forehead and crown streaked black, outlined in white. Underparts white, washed with brown on side of breast and flanks. Juvenile: spotted. Upperparts buffer than adult's; head, back and wings have pale, round buff-white spots; underparts not streaked but spotted dark brown.
DISTRIBUTION Summer visitor throughout except Iceland, in woodland edges, parks, orchards and gardens. Winters in Africa, mostly S of Equator.
HABITS AND HABITAT Often the last visitor to arrive. Population has declined since the 1960s to barely quarter of what it used to be. Obtains most of its insect food in flight, sallying forth from a perch. Nest is in a tree fork or creeper. Call: *tzee-zuk-zuk*. Song: quiet and short, like a squeaky wheelbarrow.

Red-breasted Flycatcher ■ *Ficedula parva* L 11.5cm WS 18.5–21cm

DESCRIPTION Adult male: ashy brown above, flanks washed with buff, wings and tail dark brown. Tail has diagnostic white patches on each side at base. Orange-red from chin to upper breast. Adult female and 1st-year male: similar but throat and breast buff. Juvenile: spotted, but retains white tail patches.
DISTRIBUTION Summer visitor to E Germany eastwards, and around the Baltic. Winters in India. Uncommon autumn migrant to Britain.
HABITS AND HABITAT Prefers forest with much undergrowth. Often flicks and cocks its tail. Searches canopy for insects. Secretive and difficult to find when breeding, but less so on migration. Call: harsh *zit*. Song: a cadence, descending in pitch.

male

female

Pied Flycatcher ■ *Ficedula hypoleuca* L 13cm WS 21.5–24cm

DESCRIPTION Take care to separate all Pieds from Collared (below). Breeding male: unmarked white below; black above, relieved by white forehead; white-edged tertials, which meet white wing bar; white on outer tail feathers. Non-breeding male and adult female: upperparts are brown or grey, tail and rump darker. Juvenile: looks like non-breeding adult.

DISTRIBUTION Summer visitor to W Britain, and widely from Germany eastwards into N Scandinavia and beyond. Winters in W African forests.

HABITS AND HABITAT Readily uses nestboxes. Widespread in deciduous and mixed open woodland. Eats a variety of insects. Call: loud *whit* or *wee-tic*. Song: rapid sequence of high and low notes.

FAR LEFT: *female*; LEFT: *male*

Collared Flycatcher ■ *Ficedula albicollis* L 13cm WS 22.5–24.5cm

DESCRIPTION Breeding male: unmarked white below; black above, relieved by white forehead and broad white collar, this extending right round neck; white wing panel; whitish patch on lower back and rump; black (or only mottled white) outer tail feathers. Female, immature and winter male: hard to separate from other 'brown pieds'.

DISTRIBUTION Scattered, mostly from E France and S Germany eastwards into Russia to 50–55°E and 55°N. Summer visitor, wintering S of Equator in E Africa. Vagrant to British Isles.

HABITS AND HABITAT Sunny deciduous woodland, well-timbered parks and orchards. Takes readily to nestboxes. More shy than Pied (above), and forages more in the tree canopy. The best chance of finding one is to be introduced to a nestbox population. Call: *seep* or *tick-seep*. Song: like Pied's but slower and with longer, clearer notes.

male   *female*

Great Tit ■ *Parus major*
L 14cm WS 22.5–25.5cm

DESCRIPTION Distinctive black and white head pattern; black centre stripe on yellow breast and belly; greenish back; blue-grey wings and tail, with 1 white wing bar and white outer tail.
DISTRIBUTION The most widespread tit in the world, from the British Isles in the W to Japan in the E, and from Lapland in the N to Indonesia in the S. Resident and sedentary in Britain. N and E European birds migrate S or W. Shortly after breeding in territories, forms winter flocks.
HABITS AND HABITAT A bold bird, regular at bird-feeders and nestboxes.
Found mainly in woodland, parkland, orchards, hedgerows and town gardens. Eats a large variety of insects, spiders and larvae, plus seeds and fruit in winter. Call: sharp *chink*.
Male usually has 3 or 4 distinct songs based on the typical *tea-cher*.

Coal Tit ■ *Periparus ater*
L 11.5cm WS 17–21cm

DESCRIPTION Sexes similar. Diagnostic glossy black cap and large white patch on nape. White cheeks. Chin, throat and upper breast black. Underparts buff, paler towards centre. Upperparts, wings and tail olive-grey. 2 white wing bars.
DISTRIBUTION Widespread resident to about 65°N (not Iceland), generally less common than Great (above) and Blue (p. 132) tits.
HABITS AND HABITAT The tit of coniferous woods, but also occurs in other woodland; found anywhere there are firs, even cemeteries, parks and gardens in cities. Eats insects and spiders, plus seeds in autumn and winter. Readily comes to bird tables in winter; many seeds are stored in a crevice somewhere close by. More agile than other tits when foraging. Nests in hole in a tree stump, tree, wall or rock. Call: piping *tsee*. Song: loud, clear *teechu, teechu, teechu*.

Blue Tit ■ *Cyanistes caeruleus*
L 11.5cm WS 17.5–20cm

DESCRIPTION Only tit in region with a blue crown, outlined in white. Adult: sexes similar. Dark line through eye; cheeks white, outlined in black from chin. Upperparts yellowish green. Underparts sulphur-yellow. Wings and tail dark blue, 1 white wing bar. Juvenile: similar but face white, washed yellow.

DISTRIBUTION Widespread resident to *c.* 65°N (not Iceland). Most are likely to move <10km. Populations erupt in years of high numbers.

HABITS AND HABITAT Abundant in habitats with trees, even inner-city parks and gardens. Avoids conifers. Wanders outside breeding season. Feeds high in trees; a regular visitor to bird tables in winter. Readily uses nestboxes. Inquisitive and agile in order to reach food. Young are fed on defoliating caterpillars. Call: *tsee-tsee*. Alarm: *chirr-r-r*. Song: *tsee-tsee-tsee-tsuhuhuhu* tremolo.

Crested Tit

■ *Lophophanes cristatus*
L 11.5cm WS 17–20cm

DESCRIPTION Only small bird in region with a crest – backward-pointing, black and tipped white. Adult: sexes similar. Distinctive face pattern of curving black line on a white face; black line down side of neck to join black bib. Upperparts buff-brown, wings and tail grey-brown. Juvenile: has shorter crest.

DISTRIBUTION Widespread to *c.* 65°N except in Britain, where confined to N Scotland. Not Iceland.

HABITS AND HABITAT Pine forest in the N; mixed or deciduous woods elsewhere. Very sedentary and less common than other tits. Range is further restricted by its need for rotten wood in which to excavate its nest-hole. Eats insects, spiders and seeds; stores food in autumn for use in winter. Limited vocal repertoire: low-pitched, purring trill; song using repeated calls.

Willow Tit ■ *Poecile montanus*
L 11.5cm WS 17–20.5cm

DESCRIPTION Cap extends to mantle. Bib quite extensive, with poorly defined borders. Light patch on secondaries of closed wing. Scandinavian and central European birds are greyer on back and whiter on face than British birds. Always observe plumage and call notes to be sure of identification (cf. Marsh Tit, below).
DISTRIBUTION S Scotland, England, Wales and E France, eastwards to Scandinavia, Finland and Russia.
HABITS AND HABITAT Resident in mixed woodland, coniferous forest, and trees in damp lowland. In Britain, found regularly only in last. Often the commonest tit in N Europe. Noteworthy for male and female excavating their own nest-hole, low in a very soft, rotten stump. Eats invertebrates in breeding season, seeds and berries at other times. Call: *eez-eez-eez* and characteristic nasal *tchay, tchay*.

Marsh Tit ■ *Poecile palustris*
L 11.5cm WS 18–19.5cm

DESCRIPTION Sexes and ages alike. Glossy black cap, white face. Wings, tail and upperparts greyish brown. Underparts dull white, with pale buff tinge on flanks and under-tail coverts. Separable from Willow Tit (above) with difficulty – note glossy black cap, small black bib with well-defined edges, paler underparts, and quite distinct calls and song.
DISTRIBUTION England, Wales and SE Scotland, thence eastwards to the Baltic and S Scandinavia.
HABITS AND HABITAT Resident, spending all year in the same territory, in deciduous woodland, not marshes! Does not often feed at bird tables; nests in natural holes. Eats mostly insects and spiders in spring and summer, and seeds and nuts in winter. Call: *pitchoo* and nasal *ter-char-char-char*. Song: repetition of 1 note.

Siberian Tit ■ *Poecile cinctus*
L 13.5cm WS 19.5–21cm

DESCRIPTION Sexes alike. Cap and nape sooty brown. Face white. Large sooty-black bib. Upperparts warm brown. Wings dark brown, clearly edged with greyish white. Tail grey-black with dull white outer feathers. Underparts: dull white breast and belly; rusty-red sides of breast and flanks.
DISTRIBUTION Arctic Fennoscandia and Russia, then across Siberia. Resident, but nomadic outside breeding season.
HABITS AND HABITAT Lives in coniferous forest and on tree-lined riverbanks. Very confiding even at a nest site. In winter, visits the houses of foresters to feed on bird tables and even on foresters themselves to pick up crumbs. Territorial, but gregarious in winter. Eats small invertebrates and seeds, regularly storing food in crevices. Call: *sip* and *tchay*. Song: unmusical repetition, not far-carrying, unusual among the tits.

Long-tailed Tit
■ *Aegithalos caudatus*
L 14cm WS 14–19cm

DESCRIPTION Adult: head and underparts whitish, washed with pink; black stripe from bill to mantle. Upperparts, wings and tail dull black; scapulars and rump pink; 9cm-long tail with white tips and edges. Subspecies in N and E has a pure white head and white-edged wings. Juvenile: shorter and darker, with little pink.
DISTRIBUTION Widespread resident right across Europe, but not Iceland or much of N Fennoscandia.
HABITS AND HABITAT Found in deciduous woodland, and thick scrub of gorse, Bramble *Rubus fruticosus* or Sweet-briar *Rosa rubiginosa*. Breeding pair territorial, taking *c.* 3 weeks to build a domed nest of moss covered with lichen and lined with *c.* 1,000 feathers. Non-breeding flocks keep together with the help of calls. Eats bugs, plus insect eggs and larvae. Call: *tsirrrup*. Song: rapid repetition of calls.

Bearded Tit ▪ *Panurus biarmicus*
L 12.5cm WS 16–18cm

DESCRIPTION Despite its name, is not a tit but a babbler. Adult male: unmistakable, with predominantly gingery-brown plumage, grey head and striking black moustaches. Tail very long (7cm), edged white. Under-tail coverts black. Closed wing appears banded rufous, black and white. Stubby yellow bill. Adult female: lacks male's head pattern, and is duller and less russet. Juvenile: resembles adult female with black back. Also known as Bearded Reedling.
DISTRIBUTION Breeds from the Baltic and SE Russia to SE Britain, in very scattered populations in preferred habitat. Commonest in the Netherlands. Most populations are sedentary, suffering in hard winters, but high numbers result in eruptions and birds disperse widely.
HABITS AND HABITAT Found in *Phragmites* reedbeds; species' survival depends on the protection of this habitat. Very gregarious outside breeding season. Eats invertebrates and seeds. Call: explosive, metallic *ping*. Song: quiet.

Penduline Tit ▪ *Remiz pendulinus*
L 11cm WS 16–17.5cm

DESCRIPTION Adult: sexes similar. Pale grey head with black face mask. Mantle, scapulars and wing coverts chestnut. Flight feathers black, fringed with buff. Back and rump greyish; tail black with off-white margins. Underparts off-white with chestnut smudges on sides. Juvenile: lacks black face mask and has cinnamon back.
DISTRIBUTION Scattered distribution in Germany to S Baltic eastwards. A vagrant to Britain. Winters in S Europe.
HABITS AND HABITAT Needs luxuriant vegetation by water. Gregarious and active, and will allow close approach. Eats mainly larval insects and spiders, plus seeds outside breeding season. Nest is amazing: a large, hanging, domed pouch with entrance at top. It is started by the male and finished by the pair in *c.* 3 weeks. Many males and females have several mates. Call: *tseeoo* and *tsi-tsi-tsi*. Song: a trill.

Treecreeper ■ *Certhia familiaris*
L 12.5cm WS 17.5–21cm

DESCRIPTION Brown above and white below, relieved by rufous rump, white supercilium, mottled and streaked back, 2 pale wing bars, distinctive buff band across wing, and white-spotted tertials. Tail long and brown; feathers stiff and pointed. Bill quite long and gently decurved (cf. Short-toed Treecreeper, below).
DISTRIBUTION Widely distributed from E France eastwards and throughout the British Isles. Resident, but N populations move to winter within breeding range. Numbers are hard hit by prolonged frosts.
HABITS AND HABITAT Coniferous forest in Europe, but deciduous forest in British Isles. Hard to observe because of its cryptic colours and high-pitched voice. Searches for insects and spiders up 1 tree and flies to low down on another. Call: thin *tsiew*. Song: cadence lasting 2.5–3 seconds.

Short-toed Treecreeper
■ *Certhia brachydactyla* L 12.5cm WS 17–20.5cm

DESCRIPTION Very hard to distinguish from Treecreeper (above). In close, sustained observation, note: upperparts and wings duller, browner and less obviously spotted; dull brown rump; shorter, duller supercilium; breast and rest of underparts washed grey or brown, most noticeably on flanks; bill usually looks longer and more slender; different voice.
DISTRIBUTION Sedentary. Found from Channel Islands and France as far E as Poland.
HABITS AND HABITAT Tall trees in parks, avenues, orchards and forest edges. Eats insect larvae, pupae and spiders. Call: includes a diagnostic, shrill, loud *tseep* or *zeet*, like that of a Dunnock (p. 110). Song: rapid, short phrase, lasting a second or so, *teet-teet-teetero-tit*; its short duration, loudness, lower pitch and short notes clearly distinguish it from Treecreeper's.

Red-backed Shrike

■ *Lanius collurio*
L 17cm WS 24–27cm

DESCRIPTION Adult male:
blue-grey crown and nape;
chestnut back; blue-grey
rump; black tail with white
outer feathers; pinkish-white
underparts; narrow black band
across forehead, extending
into broad black mask through
eye. Brown-black wings with
chestnut margins. Black
hawk-like bill. Adult female:

male

juvenile

rufous-brown upperparts, reddish tail, pale buff supercilium, cream underparts with brown
crescent markings. Juvenile: similar to adult female.
DISTRIBUTION Summer visitor; uncommon in N France and the Low Countries, to
around the Baltic and S Scandinavia eastwards. Uncommon migrant in Britain. Winters
in E and S Africa. Has declined in many parts.
HABITS AND HABITAT Sunny, open terrain, with bushes and small trees. Eats mainly
insects, especially beetles. Extra food is often cached by impaling it on thorns and barbed
wire, known as 'shrikes' larders'. Call: harsh *chack, chack*. Song: subdued.

Great Grey Shrike

■ *Lanius excubitor*
L 24–25cm WS 30–34cm

DESCRIPTION Adult: black face mask,
wings and tail; grey upperparts; white
underparts. White supercilium; white
patch on scapulars; white-edged long tail;
narrow white bar across base of primaries.
Hooked black bill. Juvenile: brownish-
grey upperparts and brownish-white
underparts with faint brown wavy bars.
DISTRIBUTION Arctic Fennoscandia
southwards through much of region, but

populations are scattered from Germany to France. Mainly resident; N populations are
migrants, journeying as far as S Europe. Annual autumn and winter visitor to Britain.
HABITS AND HABITAT Open country with bushes and trees. Strongly territorial in
summer and winter. Eats large insects, small reptiles and, increasingly, birds and mammals
in winter, these often impaled on a thorn to be dismembered or cached. Call: harsh *sheck,
sheck*. Song: quiet warble with mimicry.

Nuthatch ■ *Sitta europaea*
L 14cm WS 22.5–27cm

DESCRIPTION Adult: sexes similar. Short tail and woodpecker-like bill. Upperparts blue-grey, cheeks and throat white; rest of underparts orange-buff, but Fennoscandia birds paler. Broad black eye-stripe. Outer tail feathers black with white sub-terminal spots. Juvenile: duller below.
DISTRIBUTION From S Norway, S Sweden and N Russia westwards to the Atlantic coast and S Britain. Not in Ireland. Sedentary.
HABITS AND HABITAT Broadleaf and mixed woodland, and gardens with mature trees. Pair lives in the same territory year-round. Eats invertebrates and seeds; regularly visits bird tables. Only bird in region that can move head-first down a tree trunk. Female only builds the nest in a tree-hole or box, plastering entrance with mud to reduce its size. Call: loud *chwit-chwit*. Song: rapid *chu-chu-chu-* and slow *pee, pee, pee*.

Magpie ■ *Pica pica*
L 44–46cm WS 52–60cm

DESCRIPTION Unmistakable, with tail comprising over half total length. Adult: sexes similar. Scapulars, outer half of wings and flanks white. The rest black with a purple and green iridescence. Distinctive pied pattern in flight. Juvenile: has shorter tail, duller plumage.
DISTRIBUTION Sedentary resident throughout Europe, except in treeless areas. Adults may spend all their lives in same territory; 1st-year birds disperse to find their own territories.
HABITS AND HABITAT Mainly a lowland bird in lightly wooded country; as numbers have increased, it has moved into suburban and urban habitats in several countries. Feeds on ground, on invertebrates in summer, vertebrates and seed in winter, and carrion and scraps. Recent research has shown the breeding success of small birds has not been harmed by increasing numbers. Common call: loud, staccato *chacker chacker chacker chacker*.

Jay ■ *Garrulus glandarius*
L 34–35cm WS 52–58cm

DESCRIPTION Our most colourful
crow. Adult: sexes alike. Pinkish-brown
body with a white rump and under-tail
coverts; black tail; white forehead and
crown, streaked black; broad black
moustachial stripe. Wings black with
a short white bar, and shiny blue bars
on shoulders. Juvenile: similar.
DISTRIBUTION Found from within
Arctic Circle to Mediterranean,
and from Atlantic coast to Japan,
in fairly dense cover of trees, usually
broadleaved. Has increased in Britain,
even moving into towns and cities.
HABITS AND HABITAT W and S

birds are sedentary, but other populations are eruptive migrants when the acorn crop fails.
Eats invertebrates, fruits and seeds, with acorns forming the staple winter diet. Call: loud,
harsh, raucous, far-carrying *skaaak skaaak*.

Nutcracker
■ *Nucifraga caryocatactes*
L 22–33cm WS 52–58cm

DESCRIPTION Mostly chocolate-
brown head and body, streaked with
white spots. Plain dark brown cap.
Wings blackish brown with small white
spots on coverts. Tail white below and
in corners. Long, pointed black bill.
DISTRIBUTION S Fennoscandia to
about 60°N, E side of Baltic, and N
and E into Russia to above 65°N;
another population found in suitable
habitat SE from Belgium and Germany.
HABITS AND HABITAT Resident in
coniferous forest. Mainly dependent on
conifer seeds, and W birds in winter on
Hazel *Corylus avellana* nuts. If N crop

fails, hundreds or thousands erupt westwards, even reaching British Isles. In many areas
starts breeding when thick snow is still on ground. Quite noisy outside breeding season,
most distinctively with a high-pitched *kraak*.

Jackdaw ▪ *Corvus monedula*
L 33–34cm WS 67–74cm

DESCRIPTION A small black crow with a short bill. Grey rear to back of head (palest among E birds), with clear-cut edges between crown and mantle. Shiny, pearly white eyes. DISTRIBUTION Mostly resident across region to 60–65°N, except thinly in Norway and none in Iceland. E birds migrate W to winter in the breeding range. HABITS AND HABITAT Breeds and winters in a wide range of habitats: old woodland, parks, coastal cliffs, quarries and urban areas, with some open ground on which to feed on invertebrates, seeds, fruit and carrion. Lifelong pairing. Nests in small colonies in tree-holes, crevices in cliffs, nestboxes, and holes in church towers and other buildings. In non-breeding season, roosts in large numbers. Call: sharp *chack*.

Chough
▪ *Pyrrhocorax pyrrhocorax*
L 39–40cm WS 73–90cm

DESCRIPTION Entire plumage glossy black with blue and purple sheen. Long, thin, decurved red bill, and red legs. Broad wings with fingered tips. DISTRIBUTION In our region, only in W France, W Britain, and W, S and N coasts of Ireland. Mainly sedentary. HABITS AND HABITAT Favours short grass on coastal cliff tops or nearby. Feeds here, digging for invertebrates in the ground or animal dung. Nests in crevices in the cliff, on ledges in a cave or sometimes in a ruined building. Typically in small feeding flocks, but more solitary for breeding. Strong flyer, often performing aerobatics in updraughts along and over cliffs. Commonest call: loud, drawn-out *cheee-ow*, almost saying its name.

Rook ▪ *Corvus frugilegus*
L 44–46cm WS 81–99cm

DESCRIPTION Adult: all black, glossed with green and purple. Sharp, pointed black bill with whitish grey at base, and steep forehead. Loose thigh feathers give it a 'baggy trousers' effect. Juvenile: black at base of bill.
DISTRIBUTION British Isles and France eastwards into Russia, plus scattered populations in S Fennoscandia. Resident in British Isles; other N and E birds migrate S and W in winter.
HABITS AND HABITAT Mostly in agricultural country with some tall trees for nesting colonially (in a 'rookery'). Gregarious outside the breeding season for feeding, roosting and on migration. Rookeries may contain hundreds of treetop nests; there may be thousands of birds in a roost, with spectacular evening flights to it. Call: *kaah*, less harsh than that of Carrion Crow (p. 142).

Raven ▪ *Corvus corax*
L 64cm WS 120–150cm

DESCRIPTION The largest crow, one-third larger than crows or Rook. All black with wedge-shaped tail, massive bill, flat head and long wings. In flight, heavy head and bill help to give it a cruciform silhouette.
DISTRIBUTION Iceland, Faeroes, W British Isles, W France and Denmark eastwards and northwards to Arctic.
HABITS AND HABITAT From sea-level to high mountains, usually avoiding forested interiors and intensively farmed land. Needs an undisturbed nest site on a sea cliff, quarry or tree. Powerful flight; soars freely and performs aerobatics (diving, flipping on its back), mainly in breeding season. Ranges widely for a variety of food, especially carrion. Common call, often in flight, is the atmospheric sound of wild country: repeated, deep *pruk* or *kronk*.

Carrion Crow

■ *Corvus corone*
L 45–47cm WS 93–204cm

DESCRIPTION All black; unlike Rook (p. 141), has no white at base of bill, less steep forehead, more square-ended tail, and less of the 'baggy trousers' look.
DISTRIBUTION Found in England, Wales and France, eastwards to Germany and Czech Republic. Hybrids occur at borders of distribution of this and Hooded Crow (below). Resident.
HABITS AND HABITAT Breeds in a wide variety of habitats – by the coast on cliffs, on arable land and grassland, and on parkland, heaths and moors, even searching for food on estuary mud, rubbish tips or in cities. Eats almost anything! Walks with a steady gait. Nests singly, usually in a tree. Call: hoarse *kraah*.

Hooded Crow

■ *Corvus cornix*
L 45–47cm WS 93–204cm

DESCRIPTION Some authorities still consider this to be a subspecies of Carrion Crow (above). The 'Hoodie' has a dirty grey body between a black head and breast, and black wings and tail. Hybrids show great variations in colour patterns between the 2 species, some largely black, others with just a little mottling on the grey.
DISTRIBUTION Resident. Breeds in Ireland, Isle of Man, NW Scotland and its islands, Faeroes, Denmark and Fennoscandia.
HABITS AND HABITAT Found in a wide variety of habitats – by the coast on cliffs, in open woodland, parkland, heaths and moors, and even towns. Omnivorous. As with its relative, is widely considered to be vermin. Nests singly, in crown of a tree. Call: hoarse *kraah*.

Starling ■ *Sturnus vulgaris*
L 21.5cm WS 37–42cm

winter

DESCRIPTION Dumpy, short-tailed bird that walks with a waddle, not the Blackbird's hop (p. 117), and has a delta-winged shape in flight. Summer adult: black with a green and purple sheen. Winter adult: covered with whitish spots. Juvenile: dull, dirty brown with whitish throat; moults to adult winter plumage, the head the last part to change. DISTRIBUTION Breeds throughout region except high Arctic. N and E populations winter in the milder W. HABITS AND HABITAT Farmland and suburbs, especially short grass such as lawns, for feeding; digs for invertebrates with pointed bill; also eats caterpillars, seeds and fruit. Nests in holes in buildings, also in trees. Usual call: grating *cherrr*. Song: prolonged series of whistles, squeaks, warbles and mimicry. Winter flocks and roosts preceded by spectacular flights, sometimes of thousands of birds.

Golden Oriole ■ *Oriolus oriolus*
L 24cm WS 44–47cm

DESCRIPTION Adult male: bright yellow body and head with black lores, black wings with short yellow bar, black tail with yellow corners. Adult female: yellowish-green body, greenish-brown wings, and brownish-black tail with yellow corners; underparts are whiter on chin to upper breast, and are streaked dull brown. Juvenile: like adult female. DISTRIBUTION Summer visitor from S Denmark and S Finland to an outpost in East Anglia; elsewhere in Britain it is an uncommon annual migrant. Winters in Africa, mostly S of Equator. HABITS AND HABITAT Secretive. Tree-loving, in parks, large gardens, copses and open woods. Nest is a hammock filled in with a soft lining, slung between tips of 2 twigs; it may be reused the following year. Call: cat-like; alarm rattle. Song: far-carrying *weela-weeo*.

House Sparrow
■ *Passer domesticus*
L 14–15cm WS 21–25.5cm

DESCRIPTION Male: most upperparts and wings chestnut, streaked black; crown and nape grey; big black bib with broken bottom edge; dirty white cheeks; grey rump, dark brown tail; 1 white wing bar; rest of underparts dull grey. Female: brown above, all underparts dull grey; olive-brown crown; pale buff supercilium, especially noticeable behind eye.
DISTRIBUTION Found throughout – even a few in Iceland and Arctic Fennoscandia. Mostly sedentary.
HABITS AND HABITAT Breeds close to farmland, towns and cities. Ground feeder on wild seeds, cereal stubble, insects and their larvae for their young, and 'bird food' and scraps at bird tables. Nests in small colonies, in holes in trees, buildings and nestboxes, or builds domed nest in a shrub or tree. Sociable, noisy, chirruping.

Tree Sparrow ■ *Passer montanus*
L 14cm WS 20–22cm

DESCRIPTION Similar to House Sparrow (above), but note red-brown cap, small black bib with sharp bottom edge, white cheeks and incomplete collar, black patch below and behind eye, and double white wing bars. Sexes alike.
DISTRIBUTION Fennoscandia to just above 60°N, eastwards to Russia and westwards to British Isles (mostly in E and central England). Mainly sedentary, but N populations sometimes erupt.
HABITS AND HABITAT Open country with mature trees, orchards, old hedgerows, pollarded willows by slow-flowing rivers, and locally in gardens. Nests in holes in trees, buildings and nestboxes. Winters on stubble, in farmyards. Less associated with Man in the W. Call: more high-pitched than that of House Sparrow, a sharp, repeated *teck*.

Chaffinch ■ *Fringilla coelebs*
L 14.5cm WS 24.5–28.5cm

DESCRIPTION Male: multi-coloured finch with
blue-grey crown and nape, black forehead, pink cheeks
and underparts, chestnut mantle, yellowish-green rump,
black wings with intricate pattern of white shoulder
and wing bar, yellow-edged secondaries and tertials,
and black tail with white outer feathers. Female:
unmarked buffish-grey head and underparts, less
bright wing marks, olive-brown mantle.
DISTRIBUTION Widespread in British Isles and
eastwards into Fennoscandia and Russia; not Iceland.
N and E birds migrate, wintering mostly in the breeding
range.
HABITS AND HABITAT Common in deciduous and
coniferous woodland, parks and gardens. Territorial
breeder; builds a beautiful cup-shaped moss nest, lined
with hair, in a bush. Gregarious in winter. Call: metallic
chink; male's spring call a monotonous, repeated *wheet*.
Song: tuneful, ending in a flourish.

TOP: *male;* ABOVE: *female*

Brambling ■ *Fringilla montifringilla*
L 14cm WS 25–26cm

DESCRIPTION Summer male: distinctive, with a glossy
black head, mantle and bill; throat, breast and shoulders
orange; rest of underparts white; 2 white wing bars on
mostly black wings; forked black tail. In flight, shows
a noticeable white rump. Winter male: black plumage
obscured by buff feather edges; bill straw-coloured with
dark tip. Female: greyish head with buffish supercilium
and blackish striped crown. Retains male's tail and
rump pattern, but wing and body colours much duller.
DISTRIBUTION Summer visitor, breeding in
Fennoscandia from 60°N northwards, and E across
Russia. Winters S to Mediterranean and W to British
Isles.
HABITS AND HABITAT Breeds in open birch
and mixed forest. Winters especially in Beech *Fagus
sylvatica* woods, where it feeds on fallen seed ('mast'),
and on stubble, often with other finches. Common
call of migrating birds attracts observer's attention:
a tinny *tsweep*.

TOP: *male, winter;* ABOVE: *female,
winter*

Linnet ■ *Carduelis cannabina*
L 13.5cm WS 21–25.5cm

DESCRIPTION Summer male: grey head with crimson forehead and whitish crescents above and below eyes; crimson breast, buff flanks, white belly; chestnut mantle and wing coverts, white edges to blackish primaries and tail (showing well in flight). Adult female: brown above, streaked darker; spotted throat; buff breast and flanks with dark brown streaks; white belly; tail and wings have diagnostic white like summer male. Juvenile and winter male: similar to adult female. DISTRIBUTION British Isles, across the Continent to Fennoscandia, around the Baltic and into Russia. In winter, N and E birds migrate S and W; W birds migrate S as far as Iberia, more in some years than others. HABITS AND HABITAT Breeds on heaths and commons with gorse, coastal scrub and young plantations. Flight call: *chichichit*. Sings a pleasant twitter of call notes and whistles from an open perch. Large winter flocks feed on stubble and waste and fallow ground.

Twite ■ *Carduelis flavirostris*
L 14cm WS 22–24cm

DESCRIPTION Like female and juvenile Linnet, but more heavily streaked blackish above and below, with unmarked buff throat and buff tone to face; bill grey in summer but yellow in winter (Linnet's is grey); pale wing patches much less obvious than Linnet's. Breeding male has pink rump. DISTRIBUTION In summer, confined to W Ireland, central England, N Scotland and Norway. N birds winter from the Low Countries and Denmark to Poland and S Sweden. HABITS AND HABITAT Unlike Linnet, breeds on treeless grassy and heather moors, hill farms, inshore islands, cliff tops, and high fjelds (plateaux) and crags in Norway. British Isles birds are resident or move to coasts; others winter around cities in Germany and on coastal mudflats with *Salicornia* around North Sea. Call: hoarse, nasal *TWE-it*, hence name! Song: trilling and buzzing.

Lesser Redpoll ■ *Carduelis cabaret*
L 11.5cm WS <22.5cm

DESCRIPTION Overall buff tone; brown above, white below, and heavily streaked darker on wings, breast and flanks; buff wing bars. Small, pointed bill; tiny black bib; pinkish-red forehead; summer males have a red breast. The very similar **Common Redpoll** *Carduelis flammea* (L 12.5 WS >21cm) was treated as a related subspecies until recently. It is similar but paler and grey, with whitish wing bars.
DISTRIBUTION Lesser breeds in British Isles and along North Sea coast, and Common in Iceland and Fennoscandia eastwards across Russia. Lessers winter mostly in British Isles, while Commons migrate S and SE, some reaching France and Britain each year.
HABITS AND HABITAT Willow, birch, alder and juniper forest, and, especially in British Isles, conifers. Both are gregarious outside breeding season. Distinctive call: twittering *chuch-uch-uch-uch*.

TOP: *Lesser Redpoll:* ABOVE: *Common Redpoll*

Goldfinch ■ *Carduelis carduelis*
L 12cm WS 21–25.5cm

DESCRIPTION Unmistakable. Adult: sexes similar. Red–white–black-striped head, the black reaching onto crown and the white joining under red chin; upperparts, breast and flanks sandy brown; rump white; slightly forked, white-tipped black tail; wings black with a broad golden-yellow band right across and white-tipped flight feathers. Noticeably pointed bill. Juvenile: greyish-buff head and body, wings and tail as adult.
DISTRIBUTION Not Iceland or N Scotland, but rest of British Isles eastwards to S Fennoscandia, Baltic states and Russia. Winters

within the breeding area, as far S as Mediterranean.
HABITS AND HABITAT Orchards, gardens and fringes of woods and commons, wherever there is a good food source of tall weeds, especially Asteraceae. Gregarious outside breeding season. Call: tri-syllabic *tswitt-witt-witt*, like tiny bells. Song: also tinkling.

Greenfinch ■ *Carduelis chloris*
L 15cm WS 24.5–27.5cm

DESCRIPTION Plump, fork-tailed finch with stout, conical bill. Adult male: olive-green above and yellowy green below, with grey patch on base of secondaries and yellow base to primaries; blackish tail with bright yellow base. Adult female: duller, browner above, and paler and greyer below. Juvenile: like female but even less green and heavily streaked above and below. DISTRIBUTION Not Iceland; across region to Fennoscandia and Russia (mostly S of 65°N). Most are migrants, travelling short distances in the main, although N birds cover 1,000km or more.
HABITS AND HABITAT Borders of woods, and seed-bearing trees in parks, gardens and cemeteries. Feeds on ground on a great variety of seeds; can even crack open Hornbeam *Carpinus betulus* nut. Male's call: drawn-out, nasal *tswee*. Flight call: *teu teu*. Circular song flight with fluttering wings.

Siskin ■ *Carduelis spinus*
L 12cm WS 20–23cm

DESCRIPTION Tit-sized finch. Adult male: yellow face (with greyish-green ear coverts), breast, rump and outer tail. Crown and chin black. Forked tail. Wing is an intricate pattern of black, bright yellow and green, with a particularly broad yellow bar across tips of coverts and bases of primaries. Adult female: patterned as male but duller, and without black cap. Juvenile: like a dull female but more heavily streaked.
DISTRIBUTION Breeds mostly in N and W British Isles, Scandinavia and around the Baltic eastwards into Russia. Winters extensively in W Europe.
HABITS AND HABITAT Breeds in coniferous and mixed forest, especially in river valleys. Feeds on tree seeds especially, searching acrobatically in birch, alder and spruce. Regular garden visitor in late winter. Call, often in flight: clear *tsüü*. Song: non-stop twitter, ending in a wheeze.

Serin ▪ *Serinus serinus*
L 11.5cm WS 20–23cm

DESCRIPTION A tiny, fork-tailed finch, with a small bill on a largish head. Male: yellow head, throat and breast, rest of underparts white with blackish-brown streaks; head has olive crown, ear coverts and upperparts, which are boldly streaked brown; light yellow rump; wings are blackish brown with yellowish tips to feathers. Female: duller, browner above, with much paler yellow on rump and supercilium.
DISTRIBUTION Breeds in our region from France to the Baltic states; rarely in S England, but seen there most springs. Most E birds winter to W and S as far as Mediterranean.
HABITS AND HABITAT Territorial in orchards, vineyards, gardens, conifers and parkland. Gregarious in winter. Feeds mostly on small weed seeds and tree buds. Distinctive flight call: high-pitched *tirrillilit*.

Bullfinch ▪ *Pyrrhula pyrrhula*
L 14.5–16.5cm WS 22–29cm

DESCRIPTION Adult male: easily identified by black cap, ash-grey mantle and wing coverts, bright red from ear coverts to belly, white under-tail, black wings with white bar, and black tail contrasting with white rump. Adult female: same pattern as male but underparts are grey, washed with brown. Juvenile: like female but without black cap.
DISTRIBUTION Throughout region except very far N and Iceland. Mostly resident; some of the most N birds migrate into S Scandinavia and central Europe.
HABITS AND HABITAT Found in mixed woods, parks, gardens and, especially in Fennoscandia, in coniferous forest. A quiet, unobtrusive species. Commonest call note: subdued *phew*, falling in pitch. Slow, low-pitched song is mostly so quiet that many birdwatchers have never heard it!

Pine Grosbeak

■ *Pinicola enucleator*
L 18.5cm WS 30.5–35cm

DESCRIPTION The largest finch in our region, with a stout, round-shaped bill. Male: black wings with 2 white wing bars and white-edged tertials; body plumage crimson-red except grey flanks, belly and under-tail, which is barred blackish. Female: same wing colours but body plumage is shades of yellow. Both sexes have grey mottling.
DISTRIBUTION Breeds through the central spine of Fennoscandia and into Russia. Most birds are resident or eruptive, rarely seen in the W.
HABITS AND HABITAT A bird of mature, undisturbed coniferous forest with a mixture of other seed trees like birch, Rowan *Sorbus aucuparia* and Juniper. Shy in breeding season, but gregarious and not retiring through rest of year, feeding in gardens, by roadsides and in towns. Commonest call: a trisyllabic whistle, *tee-tee-tew*. Song: a short, pleasant warble.

Hawfinch

■ *Coccothraustes coccothraustes*
L 18cm WS 29–33cm

DESCRIPTION Unmistakable when seen well, with a big head and huge, triangular bill, blackish with a blue-grey base in summer and pale in winter. Male: rusty-brown upperparts, orange-buff head and underparts; grey nape and collar; black lores and chin. Wings blue-black with 2 white bars and strange club-shaped extensions to inner primaries. Female: similar but duller, and with a grey patch on secondaries. Both sexes have a short, white-tipped black tail.
DISTRIBUTION Breeds in England and France eastwards to Baltic states, S Fennoscandia and into Russia. Mostly resident.
HABITS AND HABITAT Prefers lofty deciduous forest, especially oaks, Beech *Fagus sylvatica*, Hornbeam *Carpinus betulus* and fruit trees – particularly wild and orchard cherries. Strong neck muscles enable it to crack cherry stones and other nuts neatly in half to get at the kernel. Call: explosive, loud *tzic*. Song: short and simple.

Common Crossbill

■ *Loxia curvirostra*
L 16.5cm WS 27–30.5cm

DESCRIPTION Adult male: brick-red head and body, brown wings. Bill is longer than deep, with unique crossed mandibles, the upper curved more than the lower. Adult female: grey-green with diffuse streaked upperparts, and yellower rump. Immature male:

ABOVE: *female*
LEFT: *male*

often not completely red. Juvenile: like a more heavily streaked and browner female.
DISTRIBUTION Mostly Fennoscandia and Baltic states eastwards; scattered populations W to British Isles. Resident, but irruptive some years.
HABITS AND HABITAT Breeds in conifers. Main food is spruce seed. Cone's scales are opened with the bill's tips and seed extracted with the tongue, so remains are distinctive from chewed cones left by rodents. Often can be seen close up drinking from a puddle or by a stream. Sometimes nests early, even with snow lying. Call: explosive *chip chip*.

Scottish Crossbill ■ *Loxia scotica* L 16.5cm WS 27.5–31.5cm

DESCRIPTION Very similar to Common (above), but larger head and deeper bill (especially the upper mandible).
DISTRIBUTION, HABITS AND HABITAT Resident in Scots Pine *Pinus sylvestris* forest of NE Scotland. Habits as Common. Call: explosive *chup*.

Parrot Crossbill

■ *Loxia pytyopsittacus* L 17.5cm WS 30.5–33cm

DESCRIPTION Noticeably larger than Common or Scottish (above), and more bull-necked. Plumages as for Common. Separated with care by its deep, parrot-like bill: lower mandible as deep as upper; upper mandible more strongly decurved; shorter, curved tips; and little or no 'step' from base of bill to forehead.

female

DISTRIBUTION Widespread across Fennoscandia, into the Baltic states and eastwards into Russia. Many are resident, but there is some dispersal W to Denmark, and eruptions to far W, where it has stopped to breed (e.g. England).
HABITS AND HABITAT Almost totally confined to pine forest, especially Scots Pine *Pinus sylvestris*. Feeds by cutting off a twig, putting it under its foot and prising open the cone scales to get at the seed. Call: *choop choop*, deeper and louder than Common's.

female

Common Rosefinch

■ *Carpodacus erythrinus*
L 14.5–15cm WS 24–26.5cm

DESCRIPTION The size of a sparrow, but with shorter, stout bill. Can be confused in all plumages except adult male with female House Sparrow (p. 144) or Corn Bunting (p. 155). Adult male: scarlet head, breast and rump; rest of underparts whitish with faint streaks. Mantle warm brown. Wings dark brown with 2 paler wing bar. Female and immature: olive-brown above, streaked darker; buffish white below, streaked darker; black beady eye.

DISTRIBUTION Summer visitor to S Scandinavia, E Germany and eastwards to Baltic states, Finland and Russia. Has expanded W recently, and even bred in Britain and France. Winters in India.

HABITS AND HABITAT Breeds in deciduous scrub, thickets and forest edges. Forages for seeds and buds from treetops to ground. Call: soft *twee-eek*. Song, from exposed perch, rendered as 'pleased to see you'.

Snow Bunting ■ *Plectrophenax nivalis*
L 16–17cm WS 32–38cm

DESCRIPTION Summer male: white head, rump and underparts; white wings with triangular black wing-tip and bastard wing; black tail with white outer feathers; tawny tips to mantle. Winter male: tawny colour extends over crown, ear coverts and sides of breast. Female: patterned like winter male but browner, with less white on wings.

DISTRIBUTION Breeds in Scotland (<100 pairs), Iceland, Norway, far N of Sweden, Finland, Russia and high-Arctic islands. Winters on Norwegian coast S from *c.* 65°N, around all North Sea coasts, and more widely in Scotland.

HABITS AND HABITAT Breeds in treeless moors, tundra, bare mountains and N rocky coasts. Winters on rough land by coast and shoreline. Seed-eater; always terrestrial. Call: *tirrtirrir-rip*.

ABOVE LEFT: *male, summer*
LEFT: *female, winter*

Lapland Bunting ▪ *Calcarius lapponicus*
L 15–16cm WS 25.5–28cm

DESCRIPTION Summer male: black head and breast separated from chestnut nape by white supercilium, this extending to rear of ear coverts and down beside breast. Rest of underparts white. Upperparts streaked black on chestnut, with 2 sandy stripes. Summer female: crown and ear coverts mottled brown, breast and flanks streaky. Winter adult: similar to Reed Bunting (below) but note 2 white wing bars framing chestnut panel, pale stripe in centre of crown, and heavily streaked rump.

DISTRIBUTION Summer visitor to Fennoscandia and across Russia. Winters on E coast of Britain and coastal France to Denmark and Germany, but mostly in S Russia and Ukraine.

HABITS AND HABITAT Breeds in wet grassy uplands with willows, birch and scrub. Winters in the SE in steppe country, and in the W on flat coastal fields and bare cultivated ground. Common calls: on passage, *ticky-tick-teu*; when breeding, piping *tee-leu*.

TOP: *male*; ABOVE: *juvenile*

Reed Bunting ▪ *Emberiza schoeniclus* L 15–16.5cm WS 21–28cm

DESCRIPTION Adult male: black head surrounded by white collar, this joining white moustache; black bib, rest of underparts whitish with indistinct grey-streaked flanks. Upperparts mostly brown with black and buff streaks, rump grey; wings dark brown with chestnut coverts and feather edges. Adult female: brown crown and ear coverts, creamy supercilium, white moustache, black malar stripe. Rest of underparts white, streaked with dark brown. Upperparts and wings as male. Immature: similar to female.

DISTRIBUTION Occurs right across region (except Iceland) in suitable habitats. Mostly sedentary in British Isles. N and E birds migrate SW to winter in S and W Europe.

HABITS AND HABITAT Breeds in reedbeds, rushes and thick cover by lakesides or riverbanks, and willows by moorland streams. Sociable in winter on drier ground, stubble and dunes. Call: high, soft *seeoo*. Sings from an exposed perch, repeated *tweek tweek tweek tititik*.

ABOVE: *female*
LEFT: *male*

Ortolan Bunting

■ *Emberiza hortulana*
L 16–17cm WS 23–29cm

DESCRIPTION Male: greenish-grey head and breast, pale yellow moustache and throat, rest of underparts orangey brown. Yellowish-brown rump. Tail dark brown, edged white. Mantle brown, tinged chestnut and streaked with black. Female: similar to male, but head browner and streaked. Both have a noticeable yellow eye-ring.
DISTRIBUTION In W of our region, has mostly disappeared from France; a few eastwards to Germany (where it is widespread), around the Baltic and into Russia. Summer visitor, wintering S of Sahara above 5°N.
HABITS AND HABITAT A ground feeder and nester, especially in cultivated country with plenty of trees and along tree-lined farm tracks. Call: shrill *tsee-up*; from migrants, a clear *slee*. Simple song has a ringing tone.

Yellowhammer

■ *Emberiza citrinella*
L 16–16.5cm WS 23–29.5cm

DESCRIPTION Male: distinctive. Yellow head, often almost unmarked but at other times with greenish-brown crown and ear coverts; warm brown upperparts, streaked darker; chestnut rump shows well, especially in flight; yellow underparts with chestnut on sides of breast and streaky flanks; brown-black wings and tail, the latter with white outer feathers. Female: duller, browner above and with less yellow below; more heavily marked with olive on crown, ear coverts and malar stripe; underparts more heavily marked, blackish rather than chestnut; rump reddish brown.
DISTRIBUTION Found almost throughout in suitable habitats, but not high Arctic or Iceland. Mostly winters within breeding area, but sedentary in Britain and France, and E and N birds tend to disperse or migrate further.
HABITS AND HABITAT Breeds on farmland, heaths, and coastal fields and scrub. Winters on stubble, fallow land and stackyards. Call: ringing *zit*. Song is commonly rendered as 'little bit of bread and no cheese'.

Cirl Bunting ■ *Emberiza cirlus*
L 15.5cm WS 22–25.5cm

DESCRIPTION Male: similar to male
Yellowhammer (p. 154) on upperparts and wings,
but rump yellowish grey and lightly streaked;
head boldly marked, having a black eye-stripe
with yellow above and below, yellow band below
black bib, and olive breast; rest of underparts
yellow, streaked with chestnut on flanks.
Female: more greyish than male or a female
Yellowhammer, and has a duller version of male's
head pattern.
DISTRIBUTION Resident in the region in the
most N part of its S European range; greatly
reduced in recent years, so only in France and
SW England.
HABITS AND HABITAT Breeds on pasturelands
with good hedgerows and grassy downs. Call:
quiet *sip*. Song: metallic rattle.

TOP: *male*
RIGHT: *female*

Corn Bunting
■ *Miliaria calandra* L 18cm WS 26–32cm

DESCRIPTION Looks more like a lark
than other colourful buntings. Grey-
brown above with dark streaks except on
rump; no white on tail; underparts dull
white, faintly streaked and sometimes
with a blackish central spot; large
yellowish bill; bright flesh-coloured legs,
which it dangles on short flights.
DISTRIBUTION Britain (particularly S
and E England) and France to Poland
and Ukraine. Mostly resident. Marked
decline in NW Europe in the last 40
years probably due to changing farming
practices.
HABITS AND HABITAT Needs extensive open farmland such as cereal crops, with
elevated song posts. Call: abrupt *quit*. Perches upright to sing a slow series of short notes,
gaining speed and ending with 'the jangling of a small bunch of keys'.

REFERENCES AND FURTHER READING

Aulén, G. (1996). *Where to Watch Birds in Scandinavia.* Hamlyn.

British Library (2009). *Coastal Birds: a Guide to Bird Sounds of the British Coast.* Audio CD and booklet. British Library.

British Library (2009). *Countryside Birds: a Guide to Bird Sounds of the British Countryside.* Audio CD and booklet. British Library.

Cramp, S. and Perrins, C.M. (eds) (1977–94). *The Birds of the Western Palearctic.* Vols 1–9. Oxford University Press.

Gooders, J. (1994). *Where to Watch Birds in Britain and Europe.* Christopher Helm.

Harrap, S. and Redman, N. (2003). *Where to Watch Birds – Britain.* Christopher Helm.

Milne, P. and Hutchinson, C. (2009). *Where to Watch Birds – Ireland.* Christopher Helm.

Mullarney, K., Svensson, L., Zetterström, D. and Grant, P.J. (1999). *Collins Bird Guide.* HarperCollins.

Sterry, P. (2004). *Complete British Birds.* HarperCollins.

Tucker, G.M. and Heath, M.F. (1994). *Birds in Europe: their Conservation Status.* Conservation Series No. 3. BirdLife International.

Wheatley, N. (2000). *Where to Watch Birds in Europe and Russia.* Christopher Helm.

Many books that deal in detail with one family or a single species (monographs) are also available. Famous titles over the years have included the following:

Lack, D. (1946). *The Life of the Robin.* Witherby; reprinted in 1953 by Penguin.

Ratcliffe, D. (1980). *The Peregrine Falcon.* Poyser.

Summers-Smith, J.D. (1963). *The House Sparrow.* Collins.

USEFUL ADDRESSES AND CONTACTS

Each country has its own national birdwatching society, affiliated as a 'partner' to BirdLife International, the world's leading bird research and conservation organisation. In the United Kingdom, the BirdLife International partner and leading bird protection organisation is the Royal Society for the Protection of Birds. If you are interested in helping as a volunteer on a bird survey of populations, movements or ecology, contact the United Kingdom's expert in that field, the British Trust for Ornithology. The contact details of each of these organisations are given below:

BirdLife International
Wellbrook Court
Girton Road
Cambridge
CB3 0NA
www.birdlife.org

Royal Society for the Protection of Birds (RSPB)
The Lodge
Sandy
Bedfordshire
SG19 2DL
www. rspb.org.uk

British Trust for Ornithology (BTO)
The Nunnery
Thetford
Norfolk
IP24 2PU
www.bto.org

▪ INDEX ▪